Boualem Bourahla

Vibrations, Waves & Electromagnetism

Boualem Bourahla

Vibrations, Waves & Electromagnetism

Courses and corrected exercises

Noor Publishing

Imprint

Any brand names and product names mentioned in this book are subject to trademark, brand or patent protection and are trademarks or registered trademarks of their respective holders. The use of brand names, product names, common names, trade names, product descriptions etc. even without a particular marking in this work is in no way to be construed to mean that such names may be regarded as unrestricted in respect of trademark and brand protection legislation and could thus be used by anyone.

Cover image: www.ingimage.com

Publisher:
Noor Publishing
is a trademark of
Dodo Books Indian Ocean Ltd. and OmniScriptum S.R.L publishing group

120 High Road, East Finchley, London, N2 9ED, United Kingdom
Str. Armeneasca 28/1, office 1, Chisinau MD-2012, Republic of Moldova, Europe
Printed at: see last page
ISBN: 978-620-7-47926-9

Foreword

This handout of course reminders and corrected exercises of vibrations, waves & electrognetism unit, is intended for undergraduate LMD students (L2-second year) in the two domains of formation, named "matter sciences " and "sciences and technologies ».

The manuscript brings together the courses and exercises that I have given in recent semesters within the faculties: "electrical and computer engineering" and "faculty of sciences", of M. Mammeri University of Tizi-Ouzou. Its content complies with the official program of the Ministry of Higher Education and Scientific Research of the Algerian Republic. It will allow students to follow their courses in good conditions, to better understand the series of exercises, and to pass successfully their exams in the unit "vibrations, waves & electromagnetism".

At the L2 level, the cited unit consists of two main parts: the vibrations of the system around their equilibrium positions, with one or more degrees of freedom; then the propagation of these vibrations (in addition to time, there is movement in space), which constitutes the waves.

It is useful to remember that vibrations and waves are encountered in all domains of physics: oscillating electrical circuits (in electronics), and atomic vibrations (solid-state physics) and molecular vibrations (in chemistry), mechanical waves, waves seismic waves, heat waves, electromagnetic waves of various frequencies, especially in the optical field. Often, the same equations and mathematical tools are handled, but the orders of magnitude of the quantities involved are different (amplitude, temperature, potential, etc.), which leads to a wide range of frequencies (pulsations) as well as different velocities of spread.

In order to facilitate the assimilation and understanding of the course, I have gathered some notions of mathematics, which are essential for the course of physics of vibrations and waves.

In this manual, all parts relating to waves are covered and the student will use the part that completes his profile.

I recommend that students read the referenced works, as they contain other important details.

Author: Boualem BOURAHLA
Professor in the Department of Physics, Faculty of Science, M. Mammeri University of Tizi-Ouzou.

Contents

Chapter 1

Mathematical reminders

In this physics course of the vibrations & waves unit, we are led, systematically, to carry out various mathematical operations on different sizes and physical quantities in order to determine others; therefore, the mastery of certain mathematical tools is essential. In this chapter, we recall some useful concepts.

1. Vectors and scalars

There are two types of physical quantities:

Scalar quantities: they do not change with the direction.

These include temperature, mass, time, density, etc.

Vector quantities: they are defined by their intensity and their direction.

Let us quote: speed, force, quantity of movement, displacement, etc.

1.1. Direction concept

A straight line can be traveled in two opposite directions. If we choose a positive direction, we say that the line is oriented (becomes an axis).

Therefore, a directed line defines a direction.

In 2D, the direction can be defined by an angle, (for example the plane (xOy)).

In 3D, we need two (2) or three (3) angles to define any direction.

1.2. Coordinates

To define the position of any point in a given frame, we find experimentally that it is necessary and sufficient to take three reels called coordinates.

1.3. Degree of freedom (dof)

The degree of freedom of a mechanical system is the number of possible movements independent of this system. In other words, it is the minimum number of measurable quantities necessary to define the system.

2. Coordinate systems

To describe the movement of a mobile (or point M), it is necessary to identify it by its coordinates in a coordinate system. The choice of system will depend on the characteristics of the movement.

2.1. Cartesian coordinates

The vector connecting the origin of a reference O to a point M in space is called position vector.

In Cartesian coordinates, it is written as: $\overrightarrow{OM} = x\vec{i} + y\vec{j} + z\vec{k}$, (the triplet $(\vec{i}, \vec{j}, \vec{k})$ is the basis), and x is the abscissa, y the ordinate and z the coast.

If the point M undergoes an elementary displacement, the corresponding vector is: $d\overrightarrow{OM} = dx\vec{i} + dy\vec{j} + dz\vec{k}$.

The associate elementary volume is: $d^3V = dxdydz$.

2.2. Cylindrical coordinates

In this system, the position vector is expressed by

$$\overrightarrow{OM} = \overrightarrow{Om} + \overrightarrow{mM} \equiv \rho\vec{u}_\rho + z\vec{k},$$

The point M has no component according to the vector $\vec{u}_\rho$. The cylindrical basis is a mobile one so the angle θ intervenes not in the position of M with respect to the cylindrical basis, but in the position of the cylindrical basis with respect to the frame that is fixed

Relation with Cartesian coordinates $\begin{cases} x = \rho\cos(\theta) \\ y = \rho\sin(\theta) \\ z = z \end{cases}$.

Unit vectors are $\vec{u}_\rho \begin{pmatrix} \cos(\theta) \\ \sin(\theta) \\ 0 \end{pmatrix}_{cart}$, $\vec{u}_\theta \begin{pmatrix} -\sin(\theta) \\ \cos(\theta) \\ 0 \end{pmatrix}_{cart}$

with $\rho = \sqrt{x^2 + y^2}$ and $\sin(\theta) = \dfrac{y}{\sqrt{x^2 + y^2}}$.

If the point M performs an elementary displacement, the corresponding vector is $d\overrightarrow{OM} = dr\vec{u}_\rho + \rho d\theta\,\vec{u}_\theta + dz\vec{k}$.

The associate elementary volume is $d^3V = d\rho\,d\theta\,dz$.

2.3. Spherical coordinates

In this type of coordinates, the spherical basis is movable. The position vector of point M is given by $\overrightarrow{OM} = r\vec{u}_r$.

The other two coordinates appear in the positioning of the mobile basis in relation to the fixed basis.

Relationship to Cartesian coordinates $\begin{cases} x = r\sin(\theta)\cos(\phi) \\ y = r\sin(\theta)\sin(\phi) \\ z = r\cos(\theta) \end{cases}$.

If the point M performs an elementary displacement, we write

$$d\overrightarrow{OM} = dr\vec{u}_r + rd\theta\vec{u}_\theta + r\sin(\theta)d\phi\vec{u}_\phi.$$

The associate elementary volume is $d^3V = dr \times rd\theta \times r\sin(\theta)d\phi$.

3. Mathematical operators

They are introduced to write the so-called local equations in a compact way. The most used operators in wave physics are:

3.1. Gradient $(\overrightarrow{grad})$

The operator $\overrightarrow{grad}$ represents a vector operator. It is defined by the vector having the components $(\frac{\partial}{\partial x}, \frac{\partial}{\partial y}, \frac{\partial}{\partial z})$, in vectorial basis $(\vec{i}, \vec{j}, \vec{k})$.

- Computing the gradient of a scalar multivariate function $f(x,y,z)$ gives

$$\overrightarrow{grad}f = \frac{\partial f}{\partial x}\vec{i} + \frac{\partial f}{\partial y}\vec{j} + \frac{\partial f}{\partial z}\vec{k}.$$

3.2. Nabla operator $(\vec{\nabla})$

The operator $\vec{\nabla}$ is a derivation operator. It is applicable on scalar and vector functions.

a) Its action on a scalar function f
Allows to obtain a vector.

$\vec{\nabla}.f(x,y,z) = \overrightarrow{grad}f(x,y,z)$. In this case, $\vec{\nabla} \equiv \overrightarrow{grad}$.

b) Its action on a vector $\vec{V}$
Allows having either a scalar or a vector. Indeed:

$$i)\ \vec{\nabla}.\vec{V} = (\frac{\partial}{\partial x}\vec{i} + \frac{\partial}{\partial y}\vec{j} + \frac{\partial}{\partial z}\vec{k}).(V_x\vec{i} + V_y\vec{j} + V_z\vec{k})$$

$$= \frac{\partial}{\partial x}V_x + \frac{\partial}{\partial y}V_y + \frac{\partial}{\partial z}V_z \equiv div(\vec{V}).$$

The result is a scalar in this case. It is called divergence of $\vec{V}$.

$$ii)\ \vec{\nabla}\wedge\vec{V} = (\frac{\partial}{\partial x}\vec{i} + \frac{\partial}{\partial y}\vec{j} + \frac{\partial}{\partial z}\vec{k})\wedge(V_x\vec{i} + V_y\vec{j} + V_z\vec{k})$$

$$= (\frac{\partial}{\partial y}V_z - \frac{\partial}{\partial z}V_y)\vec{i} + (\frac{\partial}{\partial z}V_x - \frac{\partial}{\partial x}V_z)\vec{j} + (\frac{\partial}{\partial x}V_y - \frac{\partial}{\partial y}V_x)\vec{k} \equiv \overrightarrow{Rot}(\vec{V}),$$

The result is a vector. It is called rotational of $\vec{V}$.

3.3. Laplacien (Δ)

It applies to both types of functions (scalar and vector). It is defined by
$$\Delta = \vec{\nabla}^2 = (\frac{\partial^2}{\partial x^2}, \frac{\partial^2}{\partial y^2}, \frac{\partial^2}{\partial z^2}).$$

Examples

- Its action on a scalar function f: $\Delta f = \dfrac{\partial^2 f}{\partial x^2} + \dfrac{\partial^2 f}{\partial y^2} + \dfrac{\partial^2 f}{\partial z^2}$.

- Its action on a vector function V:

$$\Delta V = \frac{\partial^2\vec{V}}{\partial x^2} + \frac{\partial^2\vec{V}}{\partial y^2} + \frac{\partial^2\vec{V}}{\partial z^2} = \Delta V_x\vec{i} + \Delta V_y\vec{j} + \Delta V_z\vec{k}.$$

4. Trigonometric functions

Consider any angle θ.

- The two basic functions are: $\cos(\theta)$ and $\sin(\theta)$.

- The most important relationship between them is: $(\cos(\theta))^2 + (\sin(\theta))^2 = 1$.

Its variants $\begin{cases} 1 + tg^2(\theta) = \dfrac{1}{\cos^2(\theta)} \\ ctg^2(\theta) + 1 = \dfrac{1}{\sin^2(\theta)} \end{cases}$, (to get them, just take the previous

relation and divide either by $\cos^2(\theta)$, or by $\sin^2(\theta)$.

We remind that: $\cos(\alpha \mp \beta) = \cos(\alpha)\cos(\beta) \pm \sin(\alpha)\sin(\beta)$,
$$\sin(\alpha \mp \beta) = \cos(\alpha)\sin(\beta) \mp \sin(\alpha)\cos(\beta).$$

- To simplify the cosine/sine of a sum/difference of two angles, we use the trigonometric circle of unit radius.

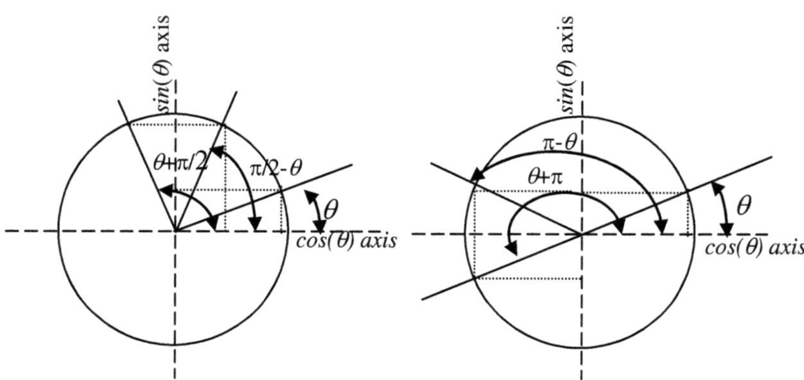

Examples
From the circles, we obtain
$\cos(-\theta) = \cos(\theta)$ and $\sin(-\theta) = -\sin(\theta)$.

$\cos(\theta + \dfrac{\pi}{2}) = -\sin(\theta)$ and $\sin(\theta + \dfrac{\pi}{2}) = \cos(\theta)$.

$\cos(\theta + \pi) = -\cos(\theta)$ and $\sin(\theta + \pi) = \sin(\theta)$.

$\sin(\dfrac{\pi}{2} - \theta) = \cos(\theta)$ and $\cos(\dfrac{\pi}{2} - \theta) = \sin(\theta)$.

$\sin(\pi - \theta) = \sin(\theta)$ and $\cos(\pi - \theta) = -\cos(\theta)$.

Derivative of trigonometric functions
To find the derivative of a trigonometric function, we go clockwise.
In general, we obtain the following results
$$\left(\sin(f(t))\right)' = f'(t)\cos(f(t)) \text{ and } \left(\cos(f(t))\right)' = -f'(t)\sin(f(t)).$$

Example
The derivative of the function $x(t) = A\cos(\omega t + \varphi)$ is
$$\frac{dx(t)}{dt} = -A\omega\sin(\omega t + \varphi).$$

5. Complex numbers
5.1. Definition
The vector space R^2 of pairs (x,y) of real numbers, can be affected by:
- an addition: $(x_1,y_1) + (x_2,y_2) = (x_1+x_2,y_1+y_2)$
- and a multiplication: $(x_1,y_1) \times (x_2,y_2) = (x_1\,x_2 - y_1\,y_2, x_1\,y_2 - x_2\,y_1)$.

These operations in R^2 is a field noted C.

The elements of C are called *complex numbers*.

Therefore: any complex number Z is written $Z = x + iy \equiv (x, y)$.

Where x represents the real part and y the imaginary part.

Proprieties
- The real number x is then identified with the pair $(x,0)$.
- The unit: $1 \equiv (1,0)$.
- The element $(0,1)$ is denoted i, with: $i^2 = (0,1) \times (0,1) = (-1,0) = -1$.

5.2. Polar representation
A point on the trigonometric circle with center 0 and radius 1, identified by a polar angle θ, has $\cos(\theta)$ as its real part and $\sin(\theta)$ as its imaginary part. The Taylor series expansion of these functions is:

$$\cos(\theta) = 1 - \frac{\theta^2}{2} + ... + (-1)^n \frac{\theta^{2n}}{2n!} + ... \qquad \text{(I)}$$

$$\sin(\theta) = \theta - \frac{\theta^3}{3!} + ... + (-1)^n \frac{\theta^{2n+1}}{(2n+1)!} + ... \qquad \text{(II)}$$

The two expansions are uniformly convergent near zero. We multiply the development of sine by the imaginary i (or j) then we do the sum, we get:

$$\cos(\theta) + i\sin(\theta) = 1 + i\theta - \frac{\theta^2}{2} - i\frac{\theta^3}{3!} ... + (-1)^n \frac{\theta^{2n}}{2n!} + i(-1)^n \frac{\theta^{2n+1}}{(2n+1)!} + ...$$

By analogy with the real exponential $e^x = \sum_{n=0}^{+\infty} \frac{x^n}{n!}$.

We pose $e^{i\theta} = \cos(\theta) + i\sin(\theta)$ (Euler's formula).

This function has the same properties as an exponential since it is defined by an identical series.

Any complex number Z can be identified, in the plane, by its polar coordinates r and θ, such that $\begin{cases} x = r\cos(\theta) \\ y = r\sin(\theta) \end{cases}$

We write $Z = x + iy = r\cos(\theta) + i\sin(\theta) = re^{i\theta}$.

The r represents the modulus of the complex number Z and θ indicates its argument.

The modulus is always positive. And the argument is defined to within $2k\pi$.

Example
We consider the complex number $Z = 1+i$.
The modulus of Z can be calculated as follows

$$r = |Z| = \sqrt{\mathrm{Re}^2 + \mathrm{Im}^2} = \sqrt{1+1} = \sqrt{2} .$$

Its argument $\cos(\theta) = \dfrac{\mathrm{Re}}{r} = \dfrac{1}{\sqrt{2}}$ and $\sin(\theta) = \dfrac{\mathrm{Im}}{r} = \dfrac{1}{\sqrt{2}}$.

$\cos(\theta) = \sin(\theta) = \dfrac{1}{\sqrt{2}} \Rightarrow \theta = \dfrac{\pi}{4} + 2k\pi$. Then, $Z = \sqrt{2}e^{i\frac{\pi}{4}}$.

6. Differential equations
6.1. Definition
An equation relating one or more dependent variables, their derivatives and their independent variables is called a differential equation.

Examples
1) $\dfrac{d^2 y}{dt^2} + xy = 0$, ($y$ dependent variable; x independent variable).

2) $\dfrac{d^2 u}{dx^2} + \dfrac{d^2 u}{dy^2} = 0$, ($u$ dependent variable; x, y independent variables).

6.2. Order of the differential equation
We call order of the differential equation, the highest order of the derivatives contained in the differential equation.

Example
$\dfrac{d^2 y}{dt^2} + 5\dfrac{dy}{dx} + 8y = 0$, is a differential equation of order 2.

An equation of order n is written $F(x, \dfrac{dy}{dt}, \dfrac{d^2 y}{dt^2} + ... + \dfrac{d^n y}{dt^n}) = 0$.

6.3. Linear and homogeneous equations

A differential equation of order n is said to be linear if it has the form

$$a_0(x)\frac{d^n y}{dt^n}+a_1(x)\frac{d^{n-1}y}{dt^{n-1}}+...+a_{n-1}(x)\frac{dy}{dt}+a_n(x)y=f(x), \text{ with } a_0(x)\neq 0.$$

$f(x)$ is called the second member of the equation.

If $f(x) = 0$, the equation is said to be homogeneous. Otherwise, the equation is said to be non-homogeneous

6.4. Typical solutions of some differential equations

6.4.a) Homogenous equations

We give some forms of equations and their general solutions.

1) $a\dfrac{dy}{dt}+by=0$ $(a\neq 0)$, its general solution has the form $y=C.e^{-\frac{b}{a}t}$.

2) $\dfrac{d^2y}{dt^2}-\lambda^2 y=0$ its solution is $y=Ae^{\lambda t}+Be^{-\lambda t}$.

3) $\dfrac{d^2y}{dt^2}+\lambda^2 y=0$ its solution is $y=Ae^{i\lambda t}+Be^{-i\lambda t}$, we can write them

also as $y=A\cos(\lambda t)+B\sin(\lambda t)$, or $y=A\cos(\lambda t-\phi)$.

4) Generalization $\dfrac{d^2y}{dt^2}+p\dfrac{dy}{dt}+qy=0$.

To find the solution, we proceed as follows:

We pose $r^2+pr+q=0$ (named characteristic equation): the solutions of the differential equation are according to the roots r_1 and r_2 of the characteristic equation.

$$\text{If }\begin{cases}\Delta\succ 0,\ y=Ae^{r_1 t}+Be^{r_2 t}\\ \Delta=0,\ y=e^{r_1 t}(At+B)\\ \Delta\prec 0,\ y=e^{\alpha t}(A\cos(\beta t)+B\sin(\beta t))\end{cases}, \begin{cases}r_1 \text{ and } r_2 \text{ reel roots}\\ a \text{ double root } r_1\\ complex\ conjugate\ roots\end{cases}$$

Note 1

In the case where $\Delta < 0$, the roots of the characteristic equation can be put in the form $r_1 = \alpha + i\beta$ et $r_2 = \alpha - i\beta)$.

6.4.b) Case where $f(x) \neq 0$

The general solution is given by $y = \overline{y} + y^*$, where $\overline{y}$ represents the solution of the homogenous differential equation (corresponding to $f(x) = 0$) and y^* refers to the particular solution of the differential equation.

Finding the particular solution

In 99 % of the cases in physics, we guess the form of a particular solution and add the coefficients it contains by identification.

That is to say, we choose a particular solution of the same form as the second member up to constants.

7. Fourier series

In analysis, Fourier series are a fundamental tool in the study of periodic functions. In technology, they are encountered in the decomposition of periodic signals, in the study of electric currents, waves, sound synthesis, image processing, etc.

Then, any non-sinusoidal periodic signal can be decomposed into a sum of trigonometric functions (Fourier series). The sum contains a constant term and sinusoidal functions of various amplitudes whose frequencies are integer multiples (known as harmonics) of the frequency of the original signal.

7.1. Reminder

1) A function f is said to be periodic, of period T, if
$\forall\, t \in \mathbb{R},\ f(t+T) = f(t)$.

2) We consider a function $f(t)$ defined on the interval $[-\pi,+\pi\,[$ (of length 2π).

This function is even if $f(t) = f(-t)$.

It is odd if $f(t) = -f(-t)$.

Note 2

Any function can be written as the sum of an even function and an odd function.

7.2. Definition of trigonometric series

A trigonometric series is a series having the following form:

$$a_0 + \sum_{n=1}^{\infty} (a_n \cos(n\omega t) + b_n \sin(n\omega t))\ .\ \text{Knowing that } \omega = \frac{2\pi}{T}\,.$$

The constants a_0, a_n and b_n (n = 1, 2, ...) are the coefficients of the trigonometric series. If the series converges, its sum is a periodic function of period 2π, $f(t) = f(t + 2\pi)$.

7.3. Development of a periodic functions (or signals) in Fourier series
Theorem

"Any function (or signal) $f(t)$, periodic of a period T, bounded, is equivalent to an infinite sum of sinusoids of frequency $f = \dfrac{n}{T}$ (n : is a number called

harmonic). $f(t) = a_0 + \sum\limits_{n=1}^{\infty}(a_n \cos(\dfrac{2\pi}{T}nt) + b_n \sin(\dfrac{2\pi}{T}nt))$ ".

The trigonometric series (which appears on the second member of the equation) is called the Fourier series of $f(t)$.

The coefficients a_n and b_n are the Fourier coefficients of the function $f(t)$.

• *Calculation of a_0*

It is assumed that the trigonometric series can be integrated term by term. This is for example the case if the numerical series formed with the coefficients of the trigonometric series converges absolutely, that is to say if the following positive numerical series converges:

$$|a_0| + |a_1| + |b_1| + ... + |a_n| + |b_n| + ...$$

The trigonometric series is then bounded and can be integrated term by

term. We can deduce $\int\limits_{0}^{T} f(t)\, dt = T\, a_0$, which provides the expression of a_0,

$$a_0 = \dfrac{1}{T}\int\limits_{0}^{T} f(t)\, dt\,.$$

• *Calculation of the other coefficients of the Fourier series*

To obtain the other coefficients of the series, we first calculate the following auxiliary integrals, in which n and k are strictly positive integers:

$$\int\limits_{0}^{T} \cos(n\omega t)\cos(k\omega t)\, dt = \begin{cases} 0, & n \neq k \\ T/2, & n = k \end{cases};$$

$$\int\limits_{0}^{T} \cos(n\omega t)\sin(k\omega t)\, dt = 0;$$

$$\int\limits_{0}^{T} \sin(nt)\sin(kt)\, dt = \begin{cases} 0, & n \neq k \\ \pi/2, & n = k \end{cases};$$

To determine a_k for a given $k > 0$, we multiply the two members of the equality of the Fourier series by $\cos(k\omega t)$:

$$f(t)\cos(k\omega t) = a_0 \cos(k\omega t) + \sum_{n=1}^{\infty}(a_n \cos(n\omega t) + b_n \sin(n\omega t))\cos(k\omega t).$$

The series of the second member can be increased and can therefore be integrated term by term.

We thus obtain the equality $\int_0^T f(t)\cos(k\omega t)dt = \dfrac{T}{2}a_k$.

Then, the expression of the coefficients a_n is $a_n = \dfrac{2}{T}\int_0^T f(t)\cos(\dfrac{2\pi}{T}nt)dt$.

Similarly, multiplying the two members of the Fourier series by $\sin(k\omega t)$ and by integrating over time t, we obtain the equality $\int_0^T f(t)\sin(k\omega t)dt = \dfrac{T}{2}b_k$.

From which we deduce an expression for the b_n , $b_n = \dfrac{2}{T}\int_0^T f(t)\sin(\dfrac{2\pi}{T}nt)dt$.

Note 3
- If the function $f(t)$ is even $\Rightarrow$ the coefficients $b_n = 0$.
- If the function $f(t)$ is odd $\Rightarrow$ the coefficients $a_n = 0$.
- The set of values of a_n and b_n constitutes the spectrum of $f(t)$.
- The terms a_0, a_n, $\cos(n\omega t)$, b_n, $\sin(n\omega t)$ are the Fourier components of the function f.
- The component corresponding to $n = 0$ (a_0) is the fundamental component. The other components ($n \geq 1$) are the harmonics. The $n = 1$ (is called 1st harmonic), the $n = 2$ (2nd harmonic), etc.

Comment
A periodic signal of frequency ν and of any shape can be obtained by adding to a sinusoid of frequency ν (known as fundamental) sinusoids whose frequencies are integer multiples of ν.
Similarly, any recurrent wave can be broken down into a sum of sinusoids (fundamental + harmonics).

7.4. Parseval's identity

It is given by the following relation $\dfrac{1}{T}\int_0^T f^2(t)dt = \dfrac{a_0^2}{4} + \sum_{n=1}^{\infty}\dfrac{a_n^2}{2} + \sum_{n=1}^{\infty}\dfrac{b_n^2}{2}$.

7.5. Sufficient conditions for a function to be developable in Fourier series

We are going to state a theorem giving sufficient conditions for the function $f(t)$ to be represented by a Fourier series.

The functions considered are piecewise monotonic and bounded on the considered interval, therefore only have first kind discontinuity points, i.e. with a limit on the right and a limit on the left.

We have the theorem: "if the periodic function $f(t)$ of period 2π is piecewise monotone and bounded on the segment $[-\pi,\pi]$, its Fourier series converges at all points. The sum of the obtained series $s(t)$ is equal to the value of the function $f(t)$ at the points of continuity".

At the points of discontinuity of $f(t)$, the sum of the series is equal to the arithmetic mean of the limits of the function on the left and on the right, so if it is a point of discontinuity of $f(t)$, we have $s(t)\big|_{t=c} = \dfrac{f(c-0)+f(c+0)}{2}$.

Exercise series n° 1 (Mathematical reminders)

Exercise 1.1

- Calculate the determinant of each of the matrices:

$$A = \begin{bmatrix} 1 & 3 & 2 \\ 2 & 1 & 3 \\ 3 & 2 & 1 \end{bmatrix} \text{ and } B = \begin{bmatrix} 1 & 2 & 0 & 2 \\ 0 & 1 & 3 & 4 \\ 5 & 2 & 1 & 0 \\ 3 & 1 & 0 & 2 \end{bmatrix}.$$

Exercise 1.2

Let consider the vectors $\vec{v} = xy^2\vec{i} + (x^2 - z^2)\vec{k}$ and $\vec{u} = 2\vec{i} + y^2\vec{j} - z\vec{k}$ and the scalar function $f(x, y, z) = 2xy + z^2$.

1) Calculate the dot product, then the vector product between the two vectors.

2) Give the:

    $i)$ expression of the gradient of the function $f(x, y, z)$,

    $ii)$ vector divergence of $\vec{v}$,

    $iii)$ rotational of the vector $\vec{v}$,

    $iv)$ Laplacian of the scalar function $f(x, y, z)$.

Exercise 1.3

1) Draw the trigonometric circle and place the following angles:

    $\pi/2$; $\pi/3$; $\pi/4$; $3\pi/4$; $3\pi/2$.

2) Using the trigonometric circle, give the value of the sine and cosine of each of these cited angles.

3) Write in simplified form the cosines or sinus of the angles using the trigonometric circle: $\cos(\pi+\varphi)$; $\cos(\pi/2-\varphi)$; $\sin(\pi/2+\varphi)$ et $\sin(\pi-\varphi)$.

Exercise 1.4

1) Recall the expressions of $\cos(a+b)$ and $\sin(a+b)$.
2) Deduce the expressions of $\cos(2a)$ et $\sin(2a)$.

Exercise 1.5

Let consider the alternating voltage $u(t) = 200\sin(200\pi t + \frac{\pi}{4})$; u is expressed in (mV) and t in (s). Determine

1) the maximum amplitude; the effective value;

2) the own pulsation, frequency and period;
3) the initial phase (or phase shift at the origin) in radians then in degrees;
4) the tension at the instants $t = 0$ (s) and $t = 1.2$ (s).

Exercise 1.6

The equation of motion of a particle along $(x'x)$-axis is of the form $x(t) = a\sin(\omega t - \alpha)$.

The time t is expressed in (s). The elongation x and the amplitude a are expressed in (cm). The period of the movement is equal to 4 (s). The trajectory is a line segment 12 (cm) long. At the origin of time, the particle, which moves in the positive direction of x, is displaced, with respect to its average position, by 3 (cm) in the negative direction.
1) What is the nature of the movement?
2) Calculate the values of the three constants a, ω and α.
3) Calculate the time after which $x = 1$ (cm) in the positive direction.
4) Calculate the value in degrees of the phase at $t = 2$ (s).
5) Plot on a graph the evolution of x as a function of time.
6) Determine the velocity of the particle at $t = 3$ (s).

Exercise 1.7

1) We consider $x(t) = C\cos(\omega t + \varphi)$
- Show, by expanding this expression, that $x(t)$ can be written as
$x(t) = A \cos(\omega t) + B \sin(\omega t)$.
2) Express C and φ as a function of A and B, then express A and B as a function of C and φ.
3) Application
 We give the following sum $x(t) = \sin(\omega t) + \cos(\omega t)$.
i) Calculate the amplitude C and the phase φ for the sum.
ii) We give the following functions

$$y_1(t) = 2\cos(5t + \frac{\pi}{6}) \text{ and } y_2(t) = 3\sin(4\pi(t - 0.125)).$$

- Write each function as sums of sine and cosine functions.

Exercise 1.8

1) Remind the expressions of $\cos(\theta)$ and $\sin(\theta)$ in complex notation.
2) Find expressions of $\cos(2\theta)$ and $\sin(2\theta)$ based on complex numbers.

3) Calculate the modules and arguments of the following complex numbers

$z_1 = j$, $z_2 = 1 + j$, $z_3 = 1 - j\sqrt{3}$ and $z_4 = \dfrac{3}{(\sqrt{3} - j)^2}$.

Exercise 1.9

Solve the following differential equations
(with the initial conditions $x(0) = 0$ and $\dot{x}(0) = 1$ (m/s)).

1) $\dfrac{d^2x}{dt^2} + 9x = 0$, 2) $\dfrac{d^2x}{dt^2} - 4x = 0$,

3) $\ddot{x} + 5\dot{x} + 4x = 0$, 4) $3\ddot{x} + 9x = 3$.

Exercise 1.10

We give periodic functions of period $T = 2\pi$ defined as follows

1) $f(t) = t$, $-\frac{T}{2} \le t \le \frac{T}{2}$

2) $\begin{cases} f(t) = -t, & -\frac{T}{2} \le t \prec 0 \\ f(t) = t, & 0 \prec t \le \frac{T}{2} \end{cases}$

3) $\begin{cases} f(t) = -1, & -\frac{T}{2} \le t \prec 0 \\ f(t) = 1, & 0 \prec t \le \frac{T}{2} \end{cases}$

4) $\begin{cases} f(t) = 0, & -\frac{T}{2} \le t \prec 0 \\ f(t) = t, & 0 \prec t \le \frac{T}{2} \end{cases}$

- Represent functions graphically and develop them in Fourier series.

Exercise 1.11

A signal generator supplies a voltage in period pulses $T = 2\ 10^{-3}$ (s) and of amplitude $V_0 = 15$ (V).

$V(t) \begin{cases} -15\ (V) & 0 \le t \le \frac{T}{2} \\ +15\ (V) & \frac{T}{2} \le t \le T \end{cases}$

1) Give the Fourier series expansion of the studied voltage.
2) Numerically determine the terms corresponding to $n \le 9$.

Answer key to exercise series n° 1

Exercise 1.1

$$\det(A) = \det\begin{vmatrix} \overset{+}{1} & \overset{-}{3} & \overset{+}{2} \\ 0 & 4 & 5 \\ 7 & 8 & 6 \end{vmatrix} = 1\times\begin{vmatrix} 4 & 5 \\ 8 & 6 \end{vmatrix} - 3\times\begin{vmatrix} 0 & 5 \\ 7 & 6 \end{vmatrix} + 2\times\begin{vmatrix} 0 & 4 \\ 7 & 8 \end{vmatrix}.$$

To calculate the determinant, we follow these steps:

1) The columns are noted alternately +, -, +

2) We calculate sub-determinants assigned coefficients that are:

- The (1) which is the intersection of the 1st row with the 1st column.
- The (-3) obtained by the intersection between the 1st row and the 2nd column, (the sign – comes from the sign of the 2nd column).
- The (2) is the intersection between the 1st row and the 3rd column.

$$\det(A) = 1\times\begin{vmatrix} 4 & 5 \\ 8 & 6 \end{vmatrix} - 3\times\begin{vmatrix} 0 & 5 \\ 7 & 6 \end{vmatrix} + 2\times\begin{vmatrix} 0 & 4 \\ 7 & 8 \end{vmatrix}$$

$$= 1\times(4\times 6 - 5\times 8) - 3\times(0\times 6 - 5\times 7) + 2\times(0\times 8 - 4\times 7)$$

$$= 24 - 40 + 105 - 56$$

$$= 33.$$

2) Case of matrix B (we proceed in the same way)

$$\det(B) = \det\begin{vmatrix} \overset{+}{1} & \overset{-}{2} & \overset{+}{0} & \overset{-}{2} \\ 0 & 1 & 3 & 4 \\ 5 & 2 & 1 & 0 \\ 3 & 1 & 0 & 2 \end{vmatrix},$$

$$\det(B) = 1\times\underbrace{\begin{vmatrix} 1 & 3 & 4 \\ 2 & 1 & 0 \\ 1 & 0 & 2 \end{vmatrix}}_{a} - 2\times\underbrace{\begin{vmatrix} 0 & 3 & 4 \\ 5 & 1 & 0 \\ 3 & 0 & 2 \end{vmatrix}}_{b} + 0\times\underbrace{\begin{vmatrix} 0 & 1 & 4 \\ 5 & 2 & 0 \\ 3 & 1 & 2 \end{vmatrix}}_{c} - 2\times\underbrace{\begin{vmatrix} 0 & 1 & 3 \\ 5 & 2 & 1 \\ 3 & 1 & 0 \end{vmatrix}}_{d},$$

$$\Rightarrow \det(B) = 1\times\det(a) - 2\times\det(b) + 0\times\det(c) - 2\times\det(d).$$

Then, we calculate the determinants (3×3) contained in the expression of $\det(B)$, (following the same steps as A).

$$\det(a) = \begin{vmatrix} \overset{+}{1} & \overset{-}{3} & \overset{+}{4} \\ 2 & 1 & 0 \\ 1 & 0 & 2 \end{vmatrix} = 1 \times \begin{vmatrix} 2 & 1 \\ 1 & 0 \end{vmatrix} - 3 \times \begin{vmatrix} 2 & 0 \\ 1 & 2 \end{vmatrix} + 4 \times \begin{vmatrix} 2 & 1 \\ 1 & 0 \end{vmatrix} = -14,$$

$$\det(b) = \begin{vmatrix} \overset{+}{0} & \overset{-}{3} & \overset{+}{4} \\ 5 & 1 & 0 \\ 3 & 0 & 2 \end{vmatrix} = 0 \times \begin{vmatrix} 1 & 0 \\ 0 & 2 \end{vmatrix} - 3 \times \begin{vmatrix} 5 & 0 \\ 3 & 2 \end{vmatrix} + 4 \times \begin{vmatrix} 5 & 1 \\ 3 & 0 \end{vmatrix} = -42,$$

$$\det(c) = \begin{vmatrix} \overset{+}{0} & \overset{-}{1} & \overset{+}{4} \\ 5 & 2 & 0 \\ 3 & 1 & 2 \end{vmatrix} = 0 \times \begin{vmatrix} 2 & 0 \\ 1 & 2 \end{vmatrix} - 1 \times \begin{vmatrix} 5 & 0 \\ 3 & 2 \end{vmatrix} + 4 \times \begin{vmatrix} 5 & 2 \\ 3 & 1 \end{vmatrix} = -14,$$

$$\det(d) = \begin{vmatrix} \overset{+}{0} & \overset{-}{1} & \overset{+}{3} \\ 5 & 2 & 1 \\ 3 & 1 & 0 \end{vmatrix} = 0 \times \begin{vmatrix} 2 & 1 \\ 1 & 0 \end{vmatrix} - 1 \times \begin{vmatrix} 5 & 1 \\ 3 & 0 \end{vmatrix} + 3 \times \begin{vmatrix} 5 & 2 \\ 3 & 1 \end{vmatrix} = 0,$$

$$\det(B) = 1 \times (-14) - 2 \times (-42) + 0 \times (-14) - 2 \times (0) = 70.$$

Exercise 1.2

1) The dot product between $\vec{v}$ and $\vec{u}$ is: $\vec{v}.\vec{u} = 2xy^2 + z^3 - x^2 z$.

(The result is a scalar).

- The vector product between $\vec{v}$ and $\vec{u}$ is

$$\vec{v} \wedge \vec{u} = \begin{vmatrix} \vec{i} & \vec{j} & \vec{k} \\ xy^2 & 0 & x^2 - z^2 \\ 2 & y^2 & -z \end{vmatrix} = y^2(z^2 - x^2)\vec{i} - (xy^2 z + 2x^2 - 2z^2)\vec{j} + xy^4\vec{k} \cdot$$

(The result is a vector).

2) i) the gradient of the function f

$$\overrightarrow{grad}(f) = \vec{\nabla}.f = \left(\frac{\partial}{\partial x}\vec{i} + \frac{\partial}{\partial y}\vec{j} + \frac{\partial}{\partial z}\vec{k} \right) f = y^2 - 2z \text{ (the result is a scalar).}$$

ii) The divergence of $\vec{v}$

$$div(\vec{v}) = \vec{\nabla}.\vec{v} = \left(\frac{\partial}{\partial x}\vec{i} + \frac{\partial}{\partial y}\vec{j} + \frac{\partial}{\partial z}\vec{k} \right).(xy^2\vec{i} + (x^2 - z^2)\vec{k}),$$

$div(\vec{v}) = y^2 - 2z$ (the results is a scalar).

iii) The rotational of $\vec{v}$

$$\vec{v} \wedge \vec{\nabla} = \begin{vmatrix} \vec{i} & \vec{j} & \vec{k} \\ \dfrac{\partial}{\partial x} & \dfrac{\partial}{\partial y} & \dfrac{\partial}{\partial z} \\ xy^2 & 0 & x^2 - z^2 \end{vmatrix} = -2y\,\vec{i} - 2xy\,\vec{k} \quad \text{(the result is a vector).}$$

iv) The Laplacian of the function f

$$\Delta f = div(\overrightarrow{grad}(f)) = \vec{\nabla}^2 . f = \left(\dfrac{\partial^2}{\partial x^2}\vec{i} + \dfrac{\partial^2}{\partial y^2}\vec{j} + \dfrac{\partial^2}{\partial z^2}\vec{k} \right) f = 2, \quad \text{(scalar).}$$

Exercise 1.3

1) The angles are shown in the figure
2) the corresponding sine and cosine values

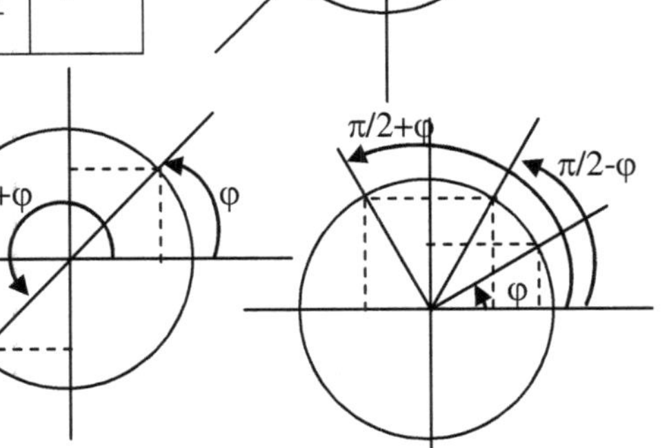

angle	$\pi/2$	$\pi/3$	$\pi/4$	$3\pi/4$	$3\pi/2$
Sinus	0	$\dfrac{1}{2}$	$\dfrac{\sqrt{2}}{2}$	$-\dfrac{\sqrt{2}}{2}$	0
cosines	1	$\dfrac{\sqrt{3}}{2}$	$\dfrac{\sqrt{2}}{2}$	$-\dfrac{\sqrt{2}}{2}$	-1

3) $\cos(\pi+\varphi) = -\cos(\varphi)$;
   $\sin(\pi+\varphi) = -\sin(\varphi)$;

   $\cos(\pi/2-\varphi) = \sin(\varphi)$;
   $\sin(\pi/2-\varphi) = \cos(\varphi)$;

   $\cos(\pi/2+\varphi) = -\sin(\varphi)$;
   $\sin(\pi/2+\varphi) = \cos(\varphi)$.

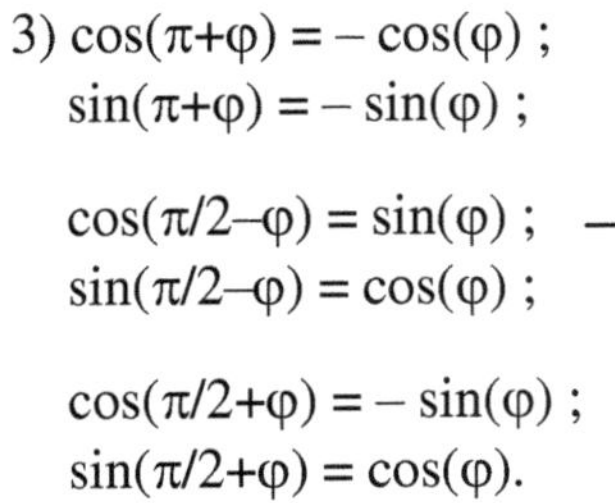

Exercise 1.4

1) We have $\cos(a+b) = \cos(a)\cos(b) - \sin(a)\sin(b)$,

and $\sin(a+b) = \cos(a)\sin(b) + \sin(a)\cos(b)$,

2) If $a = b \Rightarrow \sin(2a) = 2\cos(a)\sin(a)$.

Likewise, if $a = b \Rightarrow \cos(2a) = \cos^2(a) - \sin^2(a)$.

Exercise 1.5

1) The maximum amplitude is $u_0 = 200$ (mV).

Effective value $u_{eff} = \dfrac{u_0}{\sqrt{2}} = \dfrac{200}{\sqrt{2}} = 141.42$ (mV).

2) The own pulsation is $\omega_0 = 200\pi$ (rd/s).

The period $T_p = \dfrac{2\pi}{\omega_0} = \dfrac{2\pi}{200\pi} = 0.01$ (s)

$\Rightarrow$ the frequency $f = \dfrac{1}{T_p} = 100$ (Hz).

3) The initial phase $\phi(t=0) = \dfrac{\pi}{4}$ (rd) $= 45$ °.

4) The tension at the instants $t = 0$ (s) et $t = 1.2$ (s).

$u(0) = 200\sin(0 + \dfrac{\pi}{4}) = 141.42$ (mV) $\equiv u_{eff}$.

$u(1,2) = 200\sin(200\pi \times 1.2 + \dfrac{\pi}{4}) = 141.42$ (mV) $\equiv u_{eff}$.

Exercise 1.6

1) The movement is harmonic.

2) The amplitude $a = \dfrac{12}{2} = 6$ (cm), the period $T_p = \dfrac{2\pi}{\omega}$

$\Rightarrow \omega = \dfrac{2\pi}{T_p} = \dfrac{2\pi}{4} = 1.57$ (rd/s). We write $x(t) = 6\sin(\dfrac{\pi}{2}t - \alpha)$,

At $t = 0$, $x(0) = +3$ (cm), if we replace, we find $x(0) = 6\sin(-\alpha) = 3$.

So, $\sin(\alpha) = -\dfrac{1}{2} \Rightarrow \alpha = -\dfrac{\pi}{3}$. Finally $x(t) = 6\sin(\dfrac{\pi}{2}t + \dfrac{\pi}{3})$.

3) The time corresponding to $x = +1$ (cm).

$\Leftrightarrow 1 = 6\sin(\dfrac{\pi}{2}t + \dfrac{\pi}{3})$, this gives $\dfrac{\pi}{2}t + \dfrac{\pi}{3} = 0.0533\pi$.

$\Rightarrow t = 0.77$ (s).

4) The phase at $t = 2$ (s), $\phi = \dfrac{\pi}{2}(2) + \dfrac{\pi}{3} = \dfrac{4\pi}{3}$ (rd) $= 240°$ (or -60°).

5) The graph of $x(t)$

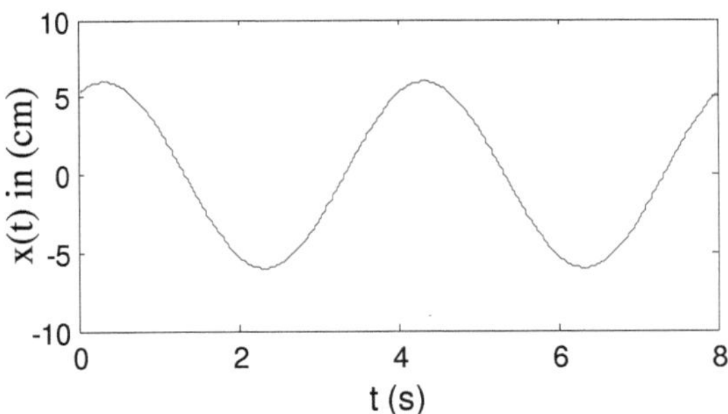

6) The velocity of the particle at $t = 3$ (s).
First, we look for the velocity by differentiating $x(t)$ with respect to time

$$v(t) = \dot{x}(t) = 3\pi \cos(\frac{\pi}{2}t + \frac{\pi}{3}) \ (\text{cm/s}).$$

Exercise 1.7

1) We know that $\cos(\alpha + \beta) = \cos(\alpha)\cos(\beta) - \sin(\alpha)\sin(\beta)$,

If we pose $\alpha = \omega t$ et $\beta = \varphi$,

we obtain $\cos(\omega t + \varphi) = \cos(\omega t)\cos(\varphi) - \sin(\omega t)\sin(\varphi)$.

$$\Rightarrow \ x(t) = C \times \cos(\omega t + \varphi) = \underbrace{C\cos(\varphi)}_{A}\cos(\omega t) - \underbrace{C\sin(\varphi)}_{B}\sin(\omega t).$$

2) According to the previous writing
$A = C\cos(\varphi)$ and $B = -C\sin(\varphi)$ (A and B as a function of C and φ).

Conversely, $A^2 + B^2 = C^2 \Rightarrow C = \sqrt{A^2 + B^2}$ and $\tan(\varphi) = \dfrac{-B}{A}$.

(Expressions of C and φ as a function of A and B).

3) Application: values of C and the phase φ in the sum of $x(t)$

i) $x(t) = \sin(\omega t) + \cos(\omega t) \Rightarrow A = 1$ and $B = 1$.

$$C = \sqrt{1^2 + 1^2} = \sqrt{2} \ \text{and} \ \tan(\varphi) = \frac{-1}{1} = -1 \Rightarrow \varphi = \frac{-\pi}{4} \ (\text{rd}).$$

ii) In the function, $y_1(t) = 2\cos(5t + \frac{\pi}{6})$, we have $C = 2$ and $\varphi = \frac{\pi}{6}$,

Which give $A = 2 \times \cos(\frac{\pi}{4}) = \sqrt{2}$ and $B = -2\sin(\frac{\pi}{4}) = -\sqrt{2}$.

Then, we write $y_1(t) = \sqrt{2}\left[\cos(5t) - \sin(5t)\right]$.

- In the same way for the function $y_2(t) = 3\sin(4\pi(t - 0.125))$.

Which give $A = 3 \times \cos(\frac{\pi}{2}) = 0$ and $B = -3\sin(\frac{\pi}{2}) = -3$.

Finally, $y_2(t) = -3\sin(4\pi t)$.

Exercise 1.8

1) We know that $e^{j\theta} = \cos(\theta) + j\sin(\theta) \Rightarrow e^{-j\theta} = \cos(\theta) - j\sin(\theta)$.

The sum of two relations gives, $\cos(\theta) = \dfrac{e^{j\theta} + e^{-j\theta}}{2}$.

The subtraction between the two relations, $\sin(\theta) = \dfrac{e^{j\theta} - e^{-j\theta}}{2j}$.

2) Expressions of $\cos(2\theta)$ and $\sin(2\theta)$ (complex notation)

We develop $\left[\cos(\theta) + j\sin(\theta)\right]^2 = \cos^2(\theta) + 2j\cos(\theta)\sin(\theta) - \sin^2(\theta)$,

According to Euler $\left[\cos(\theta) + j\sin(\theta)\right]^2 = \cos(2\theta) + j\sin(2\theta)$.

By identification, we find $\begin{cases} \cos(2\theta) = \cos^2(\theta) - \sin^2(\theta) \\ \sin(2\theta) = 2\cos(\theta)\sin(\theta). \end{cases}$

3) Modules and arguments of complex numbers

We recall that the module of a complex number $r = |z| = \sqrt{\mathrm{Re}^2\, z + \mathrm{Im}^2\, z}$.

Its argument θ is determined by searching $\cos(\theta)$ $\sin(\theta)$ by the expressions $\cos(\theta) = \dfrac{\mathrm{Re}\, z}{r}$ and $\sin(\theta) = \dfrac{\mathrm{Im}\, z}{r}$.

i) $z_1 = j \Rightarrow r_1 = |z_1| = \sqrt{0^2 + 1^2} = 1$,

The argument $\cos(\theta_1) = \dfrac{0}{1} = 0$ and $\sin(\theta_1) = \dfrac{1}{1} = 1 \Rightarrow \theta_1 = \pi/2$.

We write $z_1 = j = e^{j\frac{\pi}{2}}$.

ii) $z_2 = 1 + j \Rightarrow r_2 = |z_2| = \sqrt{1^2 + 1^2} = \sqrt{2}$,

The argument, $\cos(\theta_2) = \dfrac{1}{\sqrt{2}}$ and $\sin(\theta_2) = \dfrac{1}{\sqrt{2}} \Rightarrow \theta_2 = \pi/4$.

We write $z_2 = 1 + j = \sqrt{2}e^{j\frac{\pi}{4}}$.

iii) $z_3 = 1 - j\sqrt{3} \Rightarrow r_3 = |z_3| = \sqrt{1+3} = 2$.

The argument, $\cos(\theta_3) = \dfrac{1}{2}$ and $\sin(\theta_3) = \dfrac{-\sqrt{3}}{2} \Rightarrow \theta_3 = -\pi/3$.

We write $z_3 = 1 - j\sqrt{3} = 2e^{-j\frac{\pi}{3}}$.

iv) $z_4 = \dfrac{3}{(\sqrt{3} - j)^2} = \dfrac{3}{3 - j2\sqrt{3} - 1} = \dfrac{3}{2}\dfrac{1}{(1 - j\sqrt{3})} = \dfrac{3}{2}\dfrac{(1 + j\sqrt{3})}{(1 - j\sqrt{3})(1 + j\sqrt{3})}$.

Finally, $z_4 = \dfrac{3}{8}(1 + j\sqrt{3}) \Rightarrow r_4 = |z_4| = \dfrac{3}{4}$.

The argument, $\cos(\theta_4) = \dfrac{1}{2}$ and $\sin(\theta_4) = \dfrac{\sqrt{3}}{2} \Rightarrow \theta_4 = \pi/3$.

We write $z_4 = \dfrac{3}{4}e^{j\frac{\pi}{3}}$.

Exercise 1.9

1) The differential equation $\ddot{x} + 9x = 0$,

To solve it, we go through its characteristic equation $r^2 + 9 = 0$

($\Delta < 0 \Rightarrow$ the equation admits two complex roots $r_1 = 3j$ and $r_2 = -3j$).

Then, the solution of the differential equation is in the form

$x(t) = e^{\alpha t}\big(C_1 \cos(\beta t) + C_2 \sin(\beta t)\big)$, with $r_1 = \alpha + j\beta = 3j$ and $r_2 = \alpha - 3j$.

By replacing with the roots, we obtain $x(t) = C_1 \cos(3t) + C_2 \sin(3t)$, C_1 and C_2 are constants to be determined using the initial conditions.

At $t = 0$, $x(0) = 0$ and $\dot{x}(0) = 1$ (m/s).

$x(0) = C_1 \cos(0) + C_2 \sin(0) = 0 \Rightarrow C_1 = 0$. So, $x(t) = C_2 \sin(3t)$.

The velocity (the derivative) gives, $\dot{x}(t) = 3C_2 \cos(3t)$.

According to the initial conditions: $\dot{x}(0) = 1 = 3C_2 \cos(0) \Rightarrow C_2 = \dfrac{1}{3}$.

The final solution of the differential equation $x(t) = \dfrac{1}{3}\sin(3t)$.

2) The differential equation $\ddot{x} - 4x = 0$.

Its characteristic equation is $r^2 - 4 = 0$.

(Here, the two roots are easy to search $r_1 = 2$ and $r_2 = -2$).

We need the sign of the characteristic equation ($\Delta > 0$) to write the solution to the differential equation. It is in the form

$x(t) = C_1 e^{r_1 t} + C_2 e^{r_2 t}$, C_1 and C_2 are constants to be determined using the initial conditions. By introducing the roots $x(t) = C_1 e^{2t} + C_2 e^{-2t}$.

At $t = 0$, $x(0) = 0$ and $\dot{x}(0) = 1$ (m/s).

$x(0) = C_1 e^0 + C_2 e^0 \Rightarrow C_1 + C_2 = 0$ (I).

The derivative gives $\dot{x}(t) = 2C_1 e^{2t} - 2C_2 e^{-2t}$.

According to the initial conditions $\dot{x}(0) = 2C_1 e^0 - 2C_2 e^0 = 1$.

$\Rightarrow C_1 - C_2 = \dfrac{1}{2}$ (II).

The system composed of equations (I) and (II) gives $C_1 = C_2 = \dfrac{1}{4}$.

The final solution of the differential equation $x(t) = \dfrac{1}{4}(e^{2t} - e^{-2t})$.

3) The differential equation $\ddot{x} + 5\dot{x} + 4x = 0$.

Its characteristic equation is $r^2 + 5r + 4 = 0$.

The discriminate of the equation is $\Delta = 25 - 16 = 9 > 0$

$\Rightarrow$ the equation admits two real roots $r_1 = \dfrac{-5+3}{2} = -1$ and $r_2 = \dfrac{-5-3}{2} = -4$.

Since $\Delta > 0$, the solution of the differential equation is

$x(t) = C_1 e^{r_1 t} + C_2 e^{r_2 t}$, C_1 and C_2 are constants to be determined by using the initial conditions.

By introducing the roots $x(t) = C_1 e^{-t} + C_2 e^{-4t}$.

$x(0) = C_1 e^0 + C_2 e^0 \Rightarrow C_1 + C_2 = 0$ (I).

The derivative gives $\dot{x}(t) = -C_1 e^{-t} - 4C_2 e^{-4t}$.

From the initial conditions, $\dot{x}(0) = -C_1 e^0 - 4C_2 e^0 = 1 \Rightarrow C_1 + 4C_2 = -1$ (II).

The system composed of equations (I) and (II) gives $C_1 = -C_2 = \dfrac{1}{3}$.

The final solution of the differential equation $x(t) = \dfrac{1}{3}(e^{-t} - e^{-4t})$.

4) The differential equation $3\ddot{x} + 9x = 3$, it is equivalent to $\ddot{x} + 3x = 1$.
It is a differential equation with a right hand side.
The global solution = homogeneous solution (without 2nd member)
+ particular solution (with 2nd member).

i) Solution of the homogeneous equation $3\ddot{x}+9x=0$

Like the previous cases, the characteristic equation is $r^2+3=0$,

The two roots are $r_1=j\sqrt{3}$ and $r_2=-j\sqrt{3}$.

Here $\Delta<0$, the solution of the differential equation is of the form

$x(t)=C_1\cos(\sqrt{3}t)+C_2\sin(\sqrt{3}t)$, C_1 and C_2 are constants to be determined the initial conditions.

$x(0)=C_1\cos(0)+C_2\sin(0)=0\Rightarrow C_1=0$. Therefore, $x(t)=C_2\sin(\sqrt{3}t)$.

The derivative gives $\dot{x}(t)=\sqrt{3}C_2\cos(\sqrt{3}t)$,

From the initial conditions $\dot{x}(0)=1=\sqrt{3}C_2\cos(0)\Rightarrow C_2=\dfrac{1}{\sqrt{3}}$.

The final solution of the homogeneous equation is $x(t)=\dfrac{1}{\sqrt{3}}\sin(\sqrt{3}t)$.

ii) Particular solution of the differential equation $\ddot{x}+3x=1$

The particular solution is always of the same form as the second member.

As the 2nd member is a constant, we pose $x_p(t)=C_3$, (with C_3 constant).

Let's determine the constant C_3

x_p is a solution $\Rightarrow$ it verifies the differential equation.

We replace $0+3C_3=1\Rightarrow C_3=\dfrac{1}{3}$. Consequently, $x_p(t)=\dfrac{1}{3}$.

The overall solution of the differential equation is $x(t)=\dfrac{1}{\sqrt{3}}\sin(\sqrt{3}t)+\dfrac{1}{3}$.

Exercise 1.10

1) The function $\{f(t)=t \quad -\pi\le t\prec\pi$,

The graphic representation

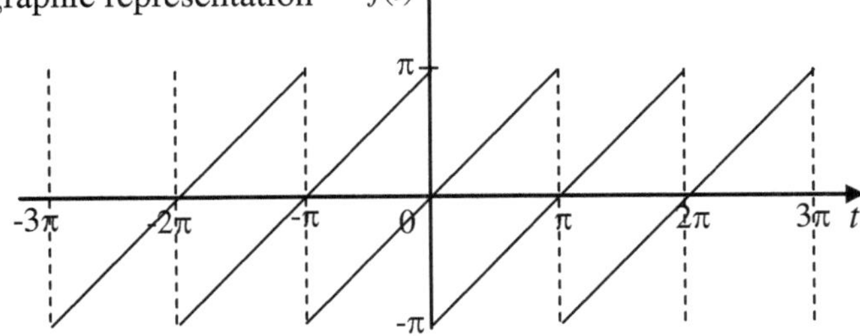

This function is piecewise monotonic and bounded.

If the function is even $\Rightarrow b_n=0$.

If the function is odd $\Rightarrow a_n = 0$.
Let's calculate its Fourier coefficients

i) $a_0 = \dfrac{1}{\pi}\left[\displaystyle\int_{-\pi}^{\pi}(t)dt\right] = 0,$

ii) $a_n = \dfrac{1}{\pi}\left[\displaystyle\int_{-\pi}^{\pi}(t)\cos(nt)dt\right] = 0,$

iii) $b_n = \dfrac{1}{\pi}\left[\displaystyle\int_{-\pi}^{\pi}(t)\sin(nt)dt\right] = \dfrac{2}{n}(-1)^{n+1}.$

The expansion thus of the function $f(t)$ in Fourier series is

$$f(t) = 2\left[\frac{\sin(t)}{1} - \frac{\sin(2t)}{2} + ... + (-1)^{p+1}\frac{\sin(pt)}{p} + ...\right].$$

2) the function $\begin{cases} f(t) = -t, & -\pi \le t \prec 0 \\ f(t) = t, & 0 \prec t \le \pi \end{cases}$,

(otherwise, $f(t) = |t|$ in the interval $[-\pi,\pi]$).
The graphic representation

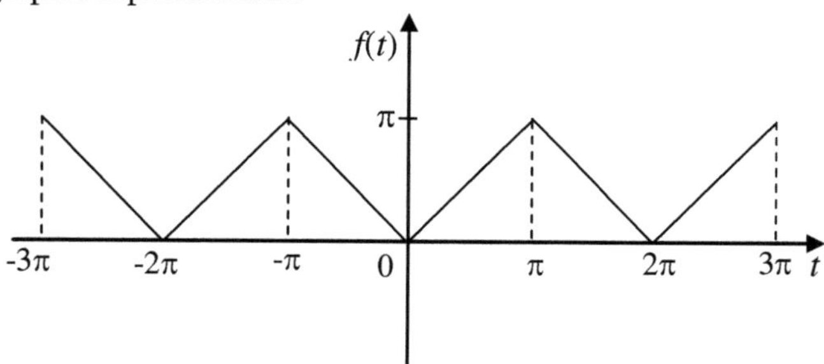

This function is piecewise monotonic and bounded. Therefore, it admits a Fourier series expansion.
Let's determine its coefficients

i) $a_0 = \dfrac{1}{\pi}\left[\displaystyle\int_{-\pi}^{0}(-t)dt + \int_{0}^{\pi} t\,dt\right] = \pi,$

ii) $a_n = \dfrac{1}{\pi}\left[\displaystyle\int_{-\pi}^{0}(-t)\cos(nt)dt + \int_{0}^{\pi}(t)\cos(nt)dt\right] = \dfrac{2}{n^2 \pi}[\cos(n\pi) - 1],$

Either $a_n = \begin{cases} 0, & n : even \\ \dfrac{-4}{n^2 \pi} & n : odd \end{cases}$

iii) $b_n = \dfrac{1}{\pi}\left[\displaystyle\int_{-\pi}^{0}(-t)\sin(nt)\,dt + \int_{0}^{\pi}(t)\sin(nt)\,dt\right] = 0,$

The Fourier series of the considered function is written

$$f(t) = \frac{\pi}{2} - \frac{4}{\pi}\left[\frac{\cos(t)}{1^2} + \frac{\cos(3t)}{3^2} + \ldots + \frac{\cos((2p+1)t)}{(2p+1)^2} + \ldots\right].$$

3) The function $\begin{cases} f(t) = -1, & -\pi \le t \prec 0 \\ f(t) = 1, & 0 \prec t \le \pi \end{cases}$

The graphic representation

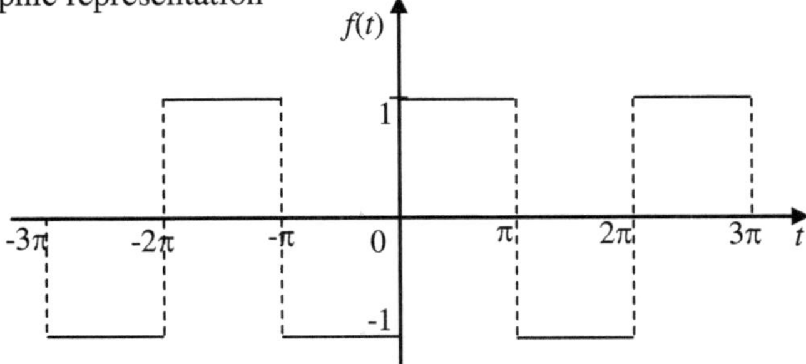

This function is piecewise monotonic and bounded.
Let's calculate its Fourier coefficients

i) $a_0 = \dfrac{1}{\pi}\left[\displaystyle\int_{-\pi}^{0}(-1)\,dt + \int_{0}^{\pi} dt\right] = 0,$

ii) $a_n = \dfrac{1}{\pi}\left[\displaystyle\int_{-\pi}^{0}(-1)\cos(nt)\,dt + \int_{0}^{\pi}\cos(nt)\,dt\right] = 0,$

iii) $b_n = \dfrac{1}{\pi}\left[\displaystyle\int_{-\pi}^{0}(-1)\sin(nt)\,dt + \int_{0}^{\pi}\sin(nt)\,dt\right] = \dfrac{2}{n\pi}(1 - \cos(n\pi)),$

Either $b_n = \begin{cases} 0, & n : even \\ \dfrac{4}{n\pi} & n : odd \end{cases}$

The Fourier series of the considered function is written

$$f(t) = \frac{4}{\pi}\left[\frac{\sin(t)}{1} + \frac{\sin(3t)}{3} + \ldots + \frac{\sin((2p+1)t)}{2p+1} + \ldots\right].$$

4) The function $\begin{cases} f(t) = 0, & -\pi \le t \prec 0 \\ f(t) = t, & 0 \prec t \le \pi \end{cases}$

The graphic representation of the function

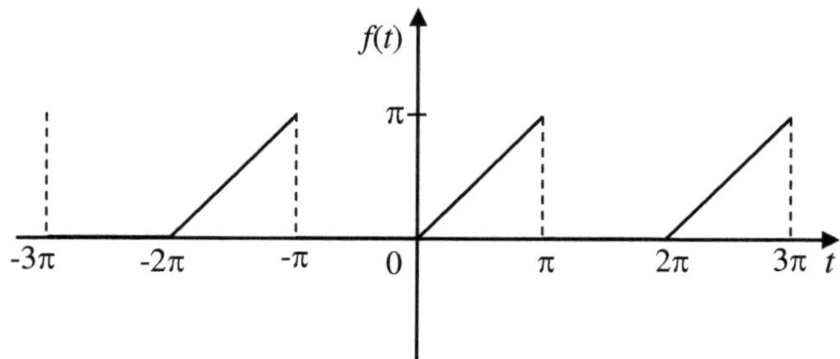

This function is piecewise monotonic and bounded.
Let's determine its coefficients

i) $a_0 = \dfrac{1}{\pi}\left[\displaystyle\int_{-\pi}^{0}(0)dt + \int_{0}^{\pi} t\,dt\right]$,

ii) $a_n = \dfrac{1}{\pi}\left[\displaystyle\int_{-\pi}^{0}(0)\cos(nt)dt + \int_{0}^{\pi}(t)\cos(nt)dt\right]$,

iii) $b_n = \dfrac{1}{\pi}\left[\displaystyle\int_{-\pi}^{0}(0)\sin(nt)dt + \int_{0}^{\pi}(t)\sin(nt)dt\right]$.

The Fourier series of the considered function is written

$$f(t) = \frac{\pi}{4} - \frac{2}{\pi}\left[\frac{\cos(t)}{1^2} + \frac{\cos(3t)}{3^2} + \frac{\cos(5t)}{5^2} + \dots - \frac{\sin(t)}{1} + \frac{\sin(2t)}{2} - \frac{\sin(3t)}{3} + \dots\right].$$

Exercise 1.11

1) The Fourier series expansion of the voltage is determined after the calculation of the following coefficients

$$a_n = \frac{1}{T}\int_{0}^{T} V(t)\cos(n\omega t)dt$$

$$= \frac{1}{T}\left[\int_{0}^{T} -V_0\,\cos(n\omega t)dt + \int_{0}^{T} V_0\,\cos(n\omega t)dt\right]$$

$$= \frac{2V_0}{n\omega T}\left(-2\sin(n\pi)\right) = 0.$$

$$b_n = \frac{1}{T}\int_{0}^{T} V(t)\sin(n\omega t)dt$$

$$= \frac{1}{T}\left[\int_{0}^{T} -V_0\,\sin(n\omega t)dt + \int_{0}^{T} V_0\,\sin(n\omega t)dt\right]$$

$$= \frac{V_0}{n\pi}\left[2 - 2(-1)^n\right].$$

For n even $\Rightarrow b_n = 0$.

For n odd $\Rightarrow b_n = \dfrac{-4V_0}{n\pi}$.

By setting $n = (2p+1)$, the development of the voltage in Fourier series is:

$$V(t) = -\frac{V_0}{n\pi} \sum_{p=0}^{\infty} \frac{\sin[(2p+1)\omega t]}{2p+1}.$$

2) Application: the terms corresponding to $n \leq 9$

b_1	b_3	b_5	b_7	b_9
-19.1	-6.37	-3.82	-2.73	-2.12

Chapter 2

Introduction to Lagrange equations

1. General information on vibrations and system dynamics

Vibrations are encountered in several areas. They are classified into several types: mechanical, elastic, acoustic, electrical, etc.

The vibration of a system constitutes a response $q(t)$ to a given excitation. It manifests itself in the form of the movement of a mobile (back and forth) around its position of equilibrium.

1.1. Some definitions

a) Vibration

Movement of a material system as a function of a time, around its equilibrium position following an excitation.

Examples

Vibration of a guitar string, swing, movement of atoms inside a solid, ...

b) Oscillation

Means periodic movement of a mobile that moves alternately on either side of a position called equilibrium called stable.

Examples

Movement of a pendulum, mass suspended from a spring moved away from its equilibrium position and released, etc.

c) Wave

It obtained when we have vibration spreading (spatial and temporal evolution of a given system vibration). It is obtained when there is propagation of a disturbance (excitation) in space in addition to its time dependence.

Examples

Mechanical waves, elastic and acoustic waves, electromagnetic waves, spin waves, etc.

1.2. Periodic movements

From the examples cited, we conclude that we can have vibrations or oscillations of physical quantities of all kinds: pressure, temperature, speed, acceleration, electric charge, electric voltage, etc.

In all cases, there is repetition of the movement in a way identical to itself at equal (regular) successive time intervals.

The duration of a repetition is called period of the movement, noted T. It is expressed in second (s), in the international system of units (SI).

Mathematically

A function $f(t)$ is said to be periodic in time when it verifies the relation: $f(t+T) = f(t)$, where T is the period of the movement.

a) Frequency is the number of repetitions of the phenomenon per unit of time is called frequency. Several notations are used. We meet N, f, n or v and its unit, in (SI), is the Hertz (Hz $\equiv$ s^{-1}).

b) Amplitude is the maximum value of the magnitude reached during the movement is called amplitude (A) or maximum elongation.

Remark

1) The main differences between vibration and oscillation are the amplitude and the frequency of the movement.

- In the example of the simple pendulum, the movement being slow, we then use the term oscillatory movement.

- The term vibration is generally reserved for higher frequencies, such as vibrations of the sound or electromagnetic vibrations, etc.

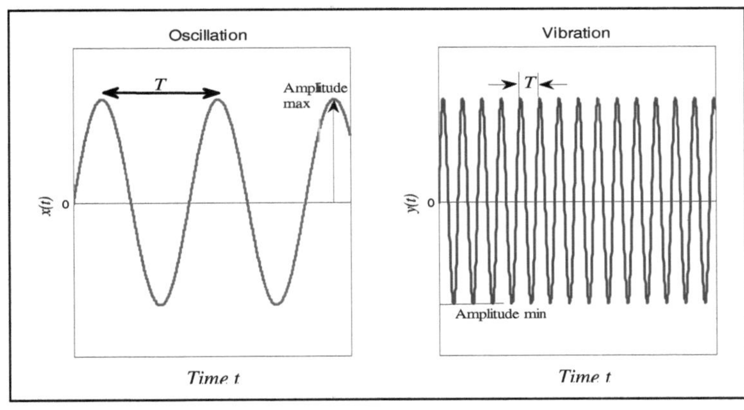

2) In one dimension, a rectilinear motion is said to be sinusoidal when the abscissa x or the elongation of the mobile is a sinusoidal function of time:
$x(t) = x_m \sin(\omega t + \varphi)$; or $x(t) = x_m \cos(\omega t + \varphi)$.
With x_m is the amplitude maximal,

$\qquad \omega$ is the movement pulsation (or frequency),

$\qquad \varphi$ is the initial phase,

$\qquad (\omega t + \varphi)$ is the phase at the time t.
In this case, we speak of harmonic oscillators.

3) Non-sinusoidal periodic phenomena (oscillation, vibration or wave) can be studied using Fourier analysis, which decomposes these periodic movements into a sum of sinusoidal functions.

4) A physical system is said to be an oscillator when the response $q(t)$ varies periodically. That's to say $q(t+T) = q(t)$, T is the temporal period of the movement.

In summary
A vibration is the repetitive movement of a physical system around its position of equilibrium. We speak of an oscillator if the repetition takes place after a regular time interval T called period, such that the response to the action verifies $q(t+T) = q(t)$.

1.3. Generalized coordinates
We call generalized coordinates the response $q(t)$ of the physical system to the excitation. They represent the independent parameters, which determine the position of the points of the system with respect to a given reference frame.
The derivative of the response (generalized coordinate) with respect to time
is noted $\dot{q}(t) = \dfrac{dq(t)}{dt}$, called generalized velocity.

In the physics course of mechanics of material point (in L1-level), to define the position of an object, in any frame, we must first choose a coordinate system. For example, the position of a mobile M in space, relative to an origin O, is defined by the position vector $\vec{r} = \overrightarrow{OM}$ at the time t.
To describe the movement of a mobile M, we use one of the three coordinate systems recalled at the beginning of this course (mathematical reminders).

$$\begin{cases} (x, y, z) & (cartesian\ coordinates) \\ (r, \theta, z) & (cylindrical\ coordinates). \\ (r, \theta, \varphi) & (spherical\ coordinates) \end{cases}$$

Knowing the position of the mobile M at the time $(t+dt)$ subsequent to t, is subordinated to knowing its velocity $\vec{v}$, given by $\vec{v} = \dfrac{d\vec{r}}{dt} = \dot{\vec{r}}$.

Result

The movement of the mobile is characterized by the function $f(t, \vec{r}, \vec{v})$, with

$\vec{r} \begin{pmatrix} x \\ y \\ z \end{pmatrix}$ is the position vector, and $\vec{v} \begin{pmatrix} \dot{x} \\ \dot{y} \\ \dot{z} \end{pmatrix}$ is the velocity vector.

Conclusions

1) In the plane, at 2D, only two (2) coordinates (x, y or r, θ) are used.

2) If the mobile M (often called mechanical system) is composed of N material points, then it will be identified by $3N$ coordinates $(x_1, y_1, z_1, x_2, y_2, z_2, ..., x_N, y_N, z_N)$.

3) It is more convenient to choose (use) for the study of a mechanical system the generalized coordinates, which are the parameters noted $q_1, q_2, q_3, ..., q_i, ..., q_n$, with $n = 3N$ independents.

We will define in the same way: the generalized velocity $\dot{q} = \dfrac{dq}{dt}$, and the generalized acceleration by $\ddot{q} = \dfrac{d\dot{q}}{dt}$.

2. Study of mechanical problems

To deal with any problem in mechanics, we use one of three formalisms, which consist in expressing the same laws of mechanics in a different way.

     i) Newton (or Newtonian's formalism),

     ii) Lagrange's formalism (known as Lagrangian),

     iii) Hamiltonian's formalism (Hamiltonian).

The choice of formalism makes it possible to find the most convenient method to solve a given problem in a simple way.

Note 1

The first two formalisms (denoted i) and ii)) are the most practical in classical mechanics, on the other hand, formalism iii) is essential for quantum mechanical problems.

In what follows, we will give an overview of each formalism, recalling the different equations that are applied.

2.1. Newtonian formalism

In this formalism, the following equations are mainly used

- Newton's equation $\vec{F} = \dfrac{d\vec{P}}{dt} = \dfrac{d(m\vec{v})}{dt} = m\vec{a}$ (m is the mass and $\vec{P}$ its amount of movement).

- Angular momentum theorem $\vec{M} = \vec{\tau} = \vec{r} \wedge \vec{F} = \vec{r} \wedge m\dfrac{d\vec{v}}{dt} = \dfrac{d}{dt}(\vec{r} \wedge \vec{P})$.

- Kinetic energy E_c given by $E_c = \dfrac{1}{2}mv^2 = \dfrac{P^2}{2m}$.

- Potential energy E_p connected to the force by the relationship

$$\vec{F} = \overrightarrow{grad}E_p = \left(\dfrac{\partial}{\partial x}\vec{i} + \dfrac{\partial}{\partial y}\vec{j} + \dfrac{\partial}{\partial z}\vec{k} \right)E_p.$$

- Mechanical energy E_m, defined as the sum of the two previous energies: $E_m = E_c + E_p$.

Example - *Case of a mass attached to a spring*

The length of the empty spring is l_0.
The length of the spring at equilibrium (the suspended mass m) is (l_0+x_0).
The length of the moving spring is (l_0+x_0+x)
The mass m is subject to its own weight and the force of the return ($\vec{p}$ and $\vec{F}$).

- A equilibrium state $\sum \vec{F} = 0 \Rightarrow \vec{p} + \vec{F_0} = 0$,

Projection $mg - kx_0 = 0$.

- Out of equilibrium situation $\sum \vec{F} = m\vec{a} \Rightarrow \vec{p} + \vec{F} = m\vec{a}$,

Projection $mg - k(x_0 + x) = m\dfrac{d^2x}{dt^2}$.

Taking into account the equilibrium condition, we get

$$\underbrace{mg - kx_0}_{=0} - kx = m\frac{d^2x}{dt^2} \;\Rightarrow\; m\ddot{x} + kx = 0 \;\text{(equation of vibrational motion)}.$$

We can put the equation in the form of the equations describing the vibrations $\ddot{x} + \omega_0^2 x = 0$, with $\omega_0^2 = \dfrac{k}{m}$

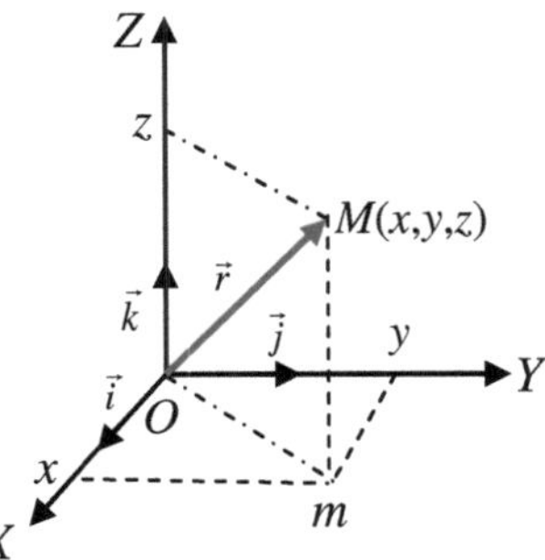

(called own pulsation, which is characteristic of the dynamics of the mechanical system).

Important
The own frequency (or pulsation) of an oscillator depends only on its parameters. It is the frequency at which an ideal free frictionless system would oscillate.

2.2. Lagrangian formalism
2.1.1. Degree of freedom (dof)
Consider the point M in space.
Its position is determined by the three components of the radius vector $\vec{r}$.
Therefore, the position of a system, composed of N material points, will be determined by $3N$ coordinates of the N ray vectors $\vec{r_i}$.
The degree of freedom, denoted dof, of a mechanical system is the number of possible movements independent of this system or it is the minimum number of measurable quantities (parameters) necessary to uniquely define the position of a physical system.

2.1.2. Lagrange's postulate
It is based on the principle of least action (called Hamilton's principle).
"The motion of any mechanical system is determined by knowing its Lagrange function $L(\dot{q}_i, q_i, t)$, with $L(\dot{q}_i, q_i, t) = E_c(\dot{q}_i, q, t) - E_p(q_i, t)$.
On each part of the trajectory, the system moves in such a way that the integral s (called action), defined by $s = \int_{t1}^{t2} L(\dot{q}_i, q_i, t)\, dt$ has the smallest possible value (min)".
This relationship leads to Lagrange's equations of motion. To show how to obtain it, we take a one-dimensional example.

2.2.3. Equation of motion of a system (at 1 dof)

Consider a system with one (1) dof $\rightarrow$ it is defined by the generalized coordinate $q(t)$, for which its action s admits a minimum.

When $q(t)$ increases by $\delta q(t)$ $\rightarrow$ the action s increases by δs, given by the expression $\delta s = \int\limits_{t1}^{t2} L(\dot{q}+\delta\dot{q}, q+\delta q, t+\delta t)\,dt - \int\limits_{t1}^{t2} L(\dot{q}, q, t)\,dt$.

The necessary condition for the action s to be minimal is that the set of terms of the first order of series expansion of δs according to the powers of $\delta\dot{q}$ and δq vanishes.

$$\delta s = \delta\int\limits_{t1}^{t2} L\,dt = 0 \Leftrightarrow \int\limits_{t1}^{t2}\left[\frac{\partial L}{\partial\dot{q}}\times\delta\left(\frac{dq}{dt}\right)+\frac{\partial L}{\partial q}\times\delta q\right]dt = 0 .$$

To calculate it, we use integration by parts
(we remind that $\int v\,du = [uv] - \int u\,dv$).

$$\Rightarrow \int\limits_{t1}^{t2}\frac{\partial L}{\partial\dot{q}}\times\delta\dot{q} = \int\limits_{t1}^{t2}\underbrace{\frac{\partial L}{\partial\dot{q}}}_{v}\times\underbrace{\left(d(\frac{\delta q}{dt})\right)}_{du} = \left[\frac{\partial L}{\partial\dot{q}}\times\delta q\right]_{t1}^{t2} - \int\limits_{t1}^{t2}\underbrace{\delta d}_{u}\times\underbrace{\left(d(\frac{\partial L}{d\dot{q}})\right)}_{dv}dt .$$

From where $s = \int\limits_{t1}^{t2}\frac{\partial L}{\partial q}\delta q\,dt + \left[\frac{\partial L}{\partial\dot{q}}\delta q\right]_{t1}^{t2} - \int\limits_{t1}^{t2}\delta q\,d\left(\frac{\partial L}{\partial\dot{q}}\right)dt = 0$,

with $\delta\big(q(t_1)\big) = \delta\big(q(t_2)\big) = 0$, we obtain

$$\delta s = \int\limits_{t1}^{t2}\left[\left(\frac{\partial L}{\partial q}\right)-\frac{d}{dt}\left(\frac{\partial L}{\partial\dot{q}}\right)\right]\delta q\,dt = 0, \forall\,\delta q .$$

That's to say $\left(\dfrac{\partial L}{\partial q}\right)-\dfrac{d}{dt}\left(\dfrac{\partial L}{\partial\dot{q}}\right) = 0$, for the coordinate q.

Reminder on the limited development of a function

Is an indispensable tool for simplification and local analysis.

In the vicinity of a stable equilibrium point, in its simplest form (this is the Newton-Leibnitz formula), we write $f'(x) = \dfrac{df(x)}{dx} = \lim\limits_{x\to x_0}\left[\dfrac{f(x)-f(x_0)}{x-x_0}\right]$,

$$\Rightarrow f(x) \approx f(x_0) + f'(x_0)(x-x_0) + O((x-x_0)^2) .$$

More generally, the successive approximations in powers of $(x\text{-}x_0)$ are given by the Taylor series

$$f(x) \approx f(x_0) + f'(x_0)(x - x_0) + \frac{1}{2} f''(x_0)(x - x_0)^2 + \dots$$

$$+ \frac{1}{n!} \frac{df}{dx}\bigg)_{x_0} (x - x_0)^n + O((x - x_0)^{n+1})$$

In particular, if x_0 is an equilibrium point, in its neighborhood, we have

$$f(x) \approx f(x_0) + \frac{1}{2} f''(x_0)(x - x_0)^2 .$$

Remarks

1) In the case of a system with n dof, we will have n equations, in the form:

$$\frac{d}{dt}\left(\frac{\partial L}{\partial \dot{q}_i}\right) - \left(\frac{\partial L}{\partial q_i}\right) = 0, \text{ with } i = 1, \dots, n.$$

These are differential equations of order 2. They are called Lagrange's equations (or equations of motion) for a conservative system (the forces derive from a potential).

2) In the case of a non-conservative system, we speak of generalized Lagrange equations,

$$\frac{d}{dt}\left(\frac{\partial L}{\partial \dot{q}_i}\right) - \left(\frac{\partial L}{\partial q_i}\right) = \sum_i F_i \text{ , (here } F_i \text{ are the applied forces).}$$

3) For more details on the demonstration of the equations of motion, you consult the works cited in references.

2.4. Examples of writing equations

a) *Case of a mass attached to a spring*

- Its kinetic energy is $E_c = \frac{1}{2} m v^2 = \frac{1}{2} m \dot{x}^2$.

- (The reference where $E_p = 0$ is taken to the plane containing the mass m in equilibrium state), then its potential energy is $E_p = \frac{1}{2} k (x - x_0)^2 - mgx$. The x_0 is the extension of the spring at equilibrium.

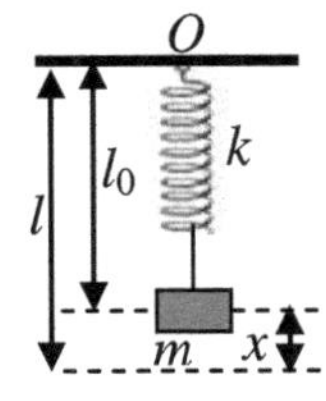

- Equilibrium condition is given by $\left.\dfrac{\partial E_p}{\partial x}\right]_{x=x_0} = 0$.

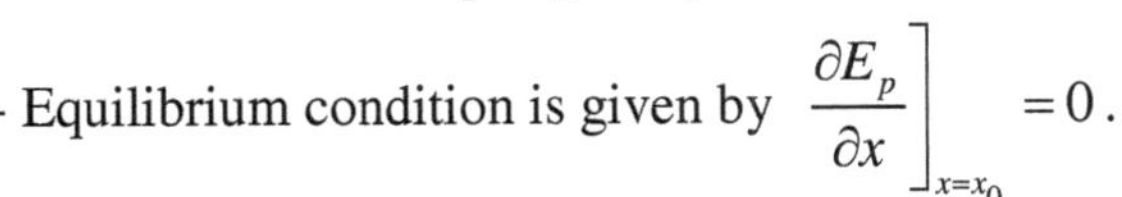

The derivative of the potential energy is $\dfrac{\partial E_p}{\partial x} = k(x - x_0) - mg$.

For $x = x_0 \Rightarrow kx_0 - mg = 0$.

Taking into account the equilibrium condition, E_p is rewritten as $E_p = \dfrac{1}{2}kx^2 + cst$.

The Lagrangian function is given by $L = E_c - E_p = \dfrac{1}{2}mv^2 - \dfrac{1}{2}kx^2 + cst$.

- The equation of motion is given by the Lagrange's equation $\dfrac{d}{dt}\left(\dfrac{\partial L}{\partial \dot{x}}\right) - \left(\dfrac{\partial L}{\partial x}\right) = 0$.

Calculating the derivatives, we obtain $\dfrac{\partial L}{\partial x} = -kx$ and $\dfrac{\partial L}{\partial \dot{q}_i} = m\dot{x}$

$$\Rightarrow \dfrac{d}{dt}\left(\dfrac{\partial L}{\partial \dot{x}}\right) = m\ddot{x} .$$

Substituting in the expression of the equations, we get $m\ddot{x} + kx = 0$.

We can put the equation in the form of the equations describing the vibrations $\ddot{x} + \omega_0^2 x = 0$, with $\omega_0^2 = \dfrac{k}{m}$ (named own pulsation).

b) *Case of a simple pendulum*

Its kinetic energy is

$$E_c = \dfrac{1}{2}mv^2 = \dfrac{1}{2}m(l\dot{\theta})^2 ,$$

- The reference $E_p = 0$ is taken at the plane containing m at equilibrium position.
Its potential energy is

$$E_p = mgh = mgl(1 - \cos(\theta))$$

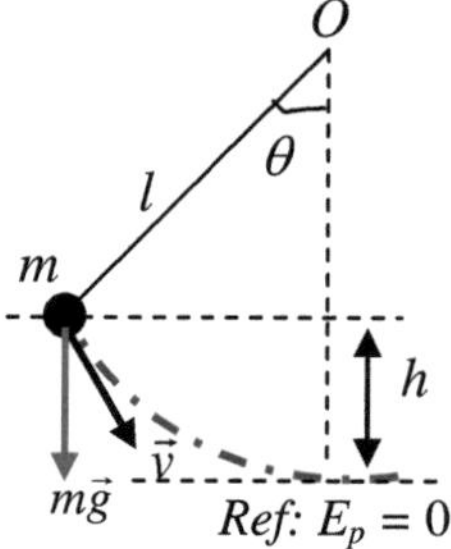

The Lagrangian is given by

$$L = E_c - E_p = \dfrac{1}{2}ml^2\dot{\theta}^2 - mgl\cos(\theta) + cst .$$

The equation of motion is given by $\dfrac{d}{dt}\left(\dfrac{\partial L}{\partial \dot{\theta}}\right) - \left(\dfrac{\partial L}{\partial \theta}\right) = 0$.

Calculating the derivatives gives $\dfrac{\partial L}{\partial \theta} = -mgl\sin(\theta)$ and $\dfrac{\partial L}{\partial \dot{\theta}} = ml^2\dot{\theta}$

$\Rightarrow \dfrac{d}{dt}\left(\dfrac{\partial L}{\dot{\theta}}\right) = ml^2\ddot{\theta}$. Substituting into the Lagrange equation, we obtain

$ml^2\ddot{\theta} + mgl\sin(\theta) = 0$.

In the case of weak oscillations $\sin(\theta) \approx \theta$, then, we write $\ddot{\theta} + \omega_0^2\theta = 0$,

with $\omega_0^2 = \dfrac{g}{l}$ (called own pulsation of the vibratory movement).

2.3. Hamiltonian's formalism

We remind that
- Newton's formalism is based on the forces.
- The Lagrange's formalism is based on the Lagrange energy function $L(q_i, \dot{q}_i, t)$.
- That of Hamilton is based on the energy of the system.

In a mechanical system of which we know the Lagrange function $L(q_i, \dot{q}_i, t)$, we choose as variables

$$\begin{cases} coordinates\ q_i,\ the\ same\ as\ those\ of\ the\ two\ others\ formalisms. \\ generalized\ impulses\ P_i,\ P_i = \dfrac{dL}{d(\dfrac{dq_i}{dt})}. \end{cases}$$

We define the function $H(q_i, \dot{q}_i, t) \rightarrow$ called the System Hamiltonian. This function represents the energy of the system. The construction of the Hamiltonian will solve problems that are purely quantum mechanical.

3. Time of oscillation

To simplify the calculations, we take a one (1) dof system and consider it in a constant external field.

In this case, the general motion of the system is described by the generalized coordinate x.

Its kinetic energy is $E_c = \dfrac{1}{2}m\dot{x}^2$. Then, the Lagrange function is given by

$$L(x, \dot{x}, t) = E_c - E_p = \dfrac{1}{2}m\dot{x}^2(t) - E_p(x, t).$$

Since the external field is constant then the total mechanical energy is conserved.

$$E_m = E_c + E_p = \frac{1}{2}m\dot{x}^2(t) + E_p(x,t) = Cst,$$

We get the expression for the velocity $\dot{x} = \dfrac{dx}{dt} = \sqrt{\dfrac{2}{m}}\sqrt{\left|E_m - E_p\right|}$

$$\Rightarrow dt = \sqrt{\frac{m}{2}}\frac{dx}{\sqrt{\left|E_m - E_p\right|}}.$$

By integration, we get the time $t = \sqrt{\dfrac{m}{2}}\displaystyle\int_{x1}^{x2}\dfrac{dx}{\sqrt{\left|E_m - E_p\right|}} + cst$.

This last relation makes it possible to calculate the period of oscillation of a vibratory movement.

$T = 2\times t.$

4. Definition of a vibration under energy aspect

We consider an example of a potential energy curve of a 1D physical system (1 dof).

Utility: to have the characteristics of the movement without making mathematical calculations.

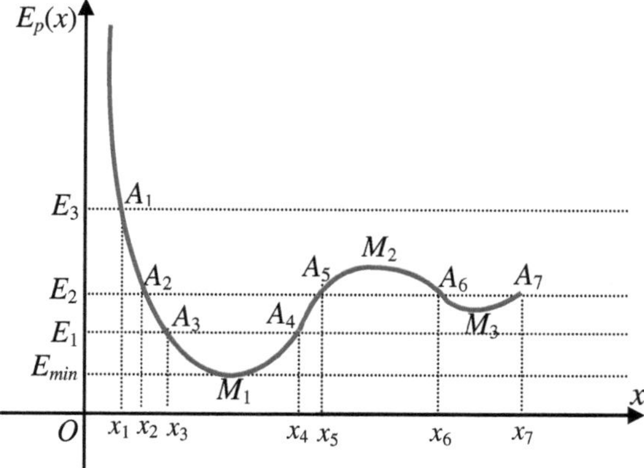

The force $F = -\dfrac{\partial E_p}{\partial x}$, $(\dfrac{\partial E_p}{\partial x}$ is the slope of the curve).

- If the slope is positive (E_p increases) $\Rightarrow F$ is negative (decreases).

- If the slope is negative (E_p decreases) $\Rightarrow F$ is positive (increases).
The extremum points are such that $F = 0$, (corresponds to the equilibrium points).
- The minima correspond to a stable equilibrium (points M_1 and M_3).
- The maximums correspond to an unstable equilibrium (point M_2).

Note 2
The first derivative allows to go back to the force. On the other hand, the second derivative gives the points of inflection (change of direction).

Discussion of the curve $E_p(x)$
    <u>Case where E_1 is the mechanical energy of the considered system</u>
Movement is possible in the interval $]x_3, x_4[$, because $E_1 > E_p$.
At the left of A_3 and the right of A_4, the potential energy $E_p > E_1 \Rightarrow E_c < 0$ (impossible ?), then points A_1 and A_2 are cusps. The movement is said to be vibratory, moreover it is oscillatory. The two points A_3 and A_4 are two stopping points, and at the level of these two points the kinetic energy is zero ($E_c = 0$).
    <u>If $E_2 > E_1$ is the mechanical energy of the system</u>
The two zones where movement is possible are $]x_2, x_5[$ et $]x_6, x_7[$,the movement is also oscillatory.
The points A_2, A_5, A_6 and A_7 are the stopped points. In the other intervals of x-axis, no movement, since $E_c = 0$.
    <u>Case where the system has total mechanical energy $E_3 > E_2$</u>
A particle coming from the right retraces the path at A_1 and goes on to infinity. In this case no vibration movement.

Conclusion
To have a vibratory movement, it is necessary:
- Two stopping points,
- $E_m > E_p$ (cuvette of potential between two points).

5. Representation of motion in phase space
5.1. Phase space
It is built from the space of positions (dim = 3) and the space of velocities (dim = 3), therefore, the space of phases is made up, in total, of dimension 6.
Whatever the configuration, it will have the dimension $2n$, with $n = 3N$, (n is the dof).

In the representation
We carry the n coordinates q_i on an axis
and the n pulses (velocities)
characterizing the system on the second axis.

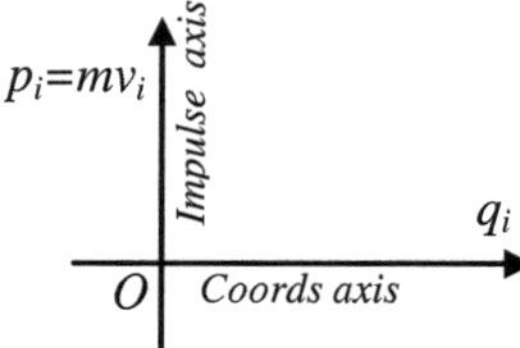

5.2. Equilibrium points

These are the points in space where the particle remains, indefinitely, in place. That is to say, the potential energy has an extremum (Max or Min).

In this case the derivative $\dfrac{\partial E_p}{\partial x}$ is zero. Likewise, the force $F = -\dfrac{\partial E_p}{\partial x} = 0$.

There are two types of equilibrium states:

a) Stable equilibrium

The curve of $E_p(x)$ has a minimum. Under an impulse pi the particle of mass m moves but goes back through the equilibrium point (tendency to return).

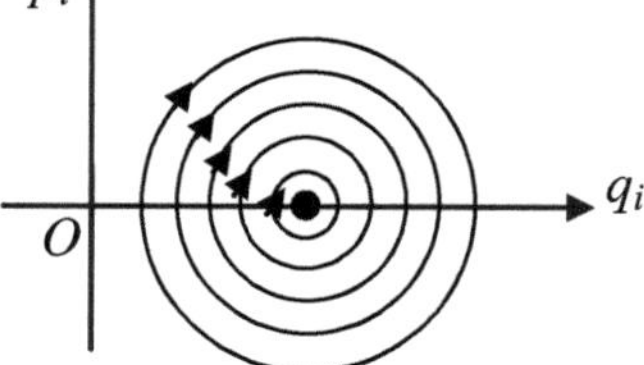

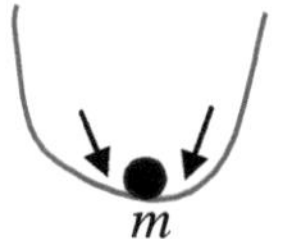

Schematic representation in phase space   Physically

b) Unstable equilibrium

The curve of $E_p(x)$ has a maximum. Under the action of an impulse p_i the particle m no longer passes through the point of equilibrium.

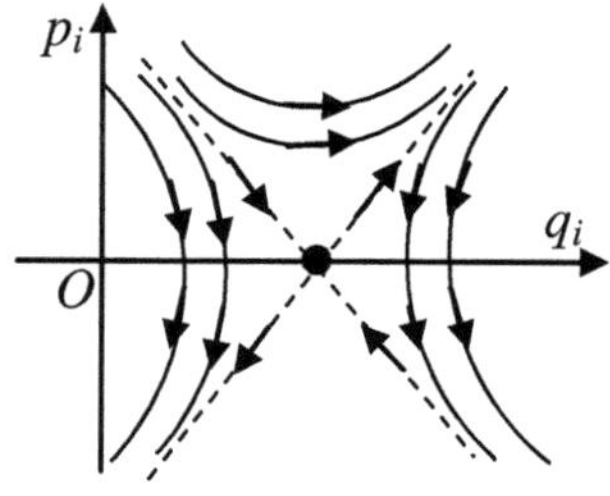

Schematic representation   in phase space   Physically

In summary

The position of stability of an oscillatory system (mechanical or electrical) is determined by calculating the following two derivatives:

- $\left(\dfrac{\partial E_p}{\partial x}\right)_{x=x_{equil}} = 0$ (inform us about the existence of the equilibrium

   position)

- $\left(\dfrac{\partial^2 E_p}{\partial x^2}\right)_{x=x_{equil}} \succ 0$ (translates that E_p is minimal for $x = x_{equil}$).

Chapter 3

Free oscillations of systems with one degree of freedom

1. Linear oscillator properties

- An oscillatory motion is said to be rectilinear with one degree of freedom (1 dof) when it takes place in a single direction in space, and the knowledge of a single position variable is sufficient to know its position.

- To define a vibration, you have to $\begin{cases} 2 \; stop. \; points \\ E_{totale} \succ E_p \end{cases}$, E_p represents a potential cuvette between these 2 stopping points around an equilibrium position x_0. And E_p has a parabolic shape. Its expression contains the term $(x - x_0)^2$.

Consider an example of a system with one degree of freedom

$$E_c = \frac{1}{2} m\dot{x}^2 = \frac{P_x^2}{2m} = E_{tot} - E_p(x) \Rightarrow P_x^2 = m\dot{x}^2 = 2m[E_{tot} - E_p(x)].$$

So, $P_x = m\dfrac{dx}{dt} = \sqrt{2m[E_{tot} - E_p(x)]}$.

Which leads to $P_x = m\dfrac{dx}{dt} = \sqrt{2m[E_{tot} - E_p(x)]}$.

2. Motion and equation of a linear oscillation

A linear vibration is characterized by the quadratic form of the potential energy $E_p(x)$.

As $E_p(x)$ is proportional to $(x - x_0)^2$, with a proportionality

coefficient $\alpha = \dfrac{1}{2} k$.

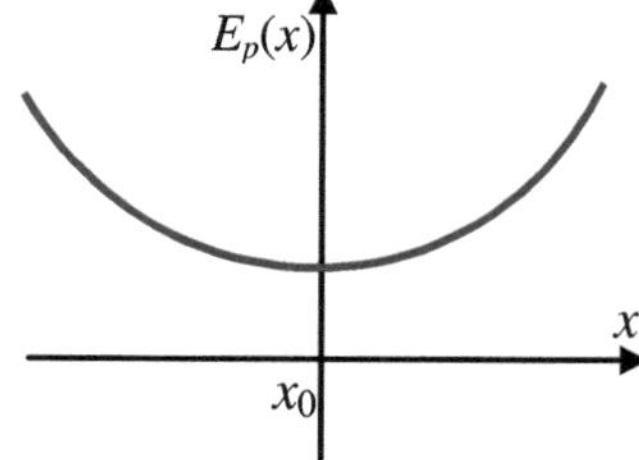

In the case where $x_0 = 0 \Rightarrow E_p(x) = \frac{1}{2} kx^2$.

The Lagrangian is then written in the form $L = \dfrac{1}{2}m\dot{x}^2 - \dfrac{1}{2}kx^2$.

The system is conservative $\Rightarrow \dfrac{d}{dt}\left(\dfrac{\partial L}{\partial \dot{x}}\right) - \dfrac{\partial L}{\partial x} = 0$.

Which give $m\ddot{x} + kx = 0$, equation governing motion.
(2nd order differential equation with constant coefficients).

The equation can be put in the following general form $\ddot{x} + \omega^2 x = 0$, with

$\omega^2 = \dfrac{k}{m}$, (named own vibration pulsation (or frequency) of the considered system).

The differential equation admits 2 independent solutions $x = \cos(\omega t)$ and $x = \sin(\omega t)$, hence the general solution

$x(t) = a\cos(\omega t) + b\sin(\omega t)$, a and b are $csts$.

The later can be written in the following general form $x(t) = A\cos(\omega t + \varphi)$,

with $A = \sqrt{a^2 + b^2}$ and $tg(\varphi) = \dfrac{-b}{a}$.

    A is called oscillation amplitude (unit in m).

    ω Pulsation (or frequency) (unit in rad/s).

    φ Phase depending on the initial conditions.

Note 1

In complex notation, $x(t) = A\cos(\omega t + \varphi) = \mathrm{Re}\left\{Ae^{j(\omega t + \varphi)}\right\}$, (notation identical to the previous one).

3. Classification of linear oscillations

We distinguish three kinds of oscillations

i) Free oscillations, in this type of oscillation, there is no energy input from the external environment. The potential energy of the system is a function of the square of the coordinate.

ii) Damped oscillations, these are oscillations that are characterized by the existence of friction forces (forces not deriving from a potential).

iii) Forced oscillations, there is a supply of energy from the external environment, in this situation.

4. Study of linear oscillations
4.1. Free oscillations

In this case, $E_c = \dfrac{1}{2}m\dot{x}^2$ and $E_p(x) = \dfrac{1}{2}kx^2$.

The Lagrangian is written as $L = \dfrac{1}{2}m\dot{x}^2 - \dfrac{1}{2}kx^2$.

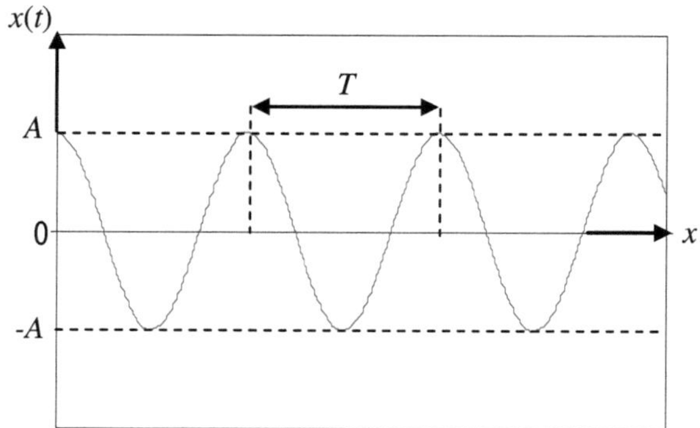

Lagrange's equations are given by $\dfrac{d}{dt}\left(\dfrac{\partial L}{\partial \dot{x}}\right) - \dfrac{\partial L}{\partial x} = 0$.

The calculation of the derivatives gives $m\ddot{x} + kx = 0$, which we will put in the form $\ddot{x} + \omega^2 x = 0$, (with $\omega^2 = \dfrac{k}{m}$).

Its solution is determined by $x(t) = A\cos(\omega t + \varphi) = \text{Re}\left\{Ae^{j(\omega t + \varphi)}\right\}$.

4.2. Damped oscillations

In this situation, the Lagrangian is written as $L = \dfrac{1}{2}m\dot{x}^2 - \dfrac{1}{2}kx^2 + xF(t)$.

Here $F(t)$ is a force not deriving from a potential, so the Lagrange equations will take the following form $\dfrac{d}{dt}\left(\dfrac{\partial L}{\partial \dot{x}}\right) - \dfrac{\partial L}{\partial x} = \sum generalized\ forces\ F(t)$.

This gives the equation of motion $m\ddot{x} + kx = F(t)$.

Suppose that the force $F(t)$ is related to viscous friction, we then define it in the form $F(t) = -\alpha\,\dot{x} = -\alpha\dfrac{dx}{dt}$ (force inversely proportional to velocity).

By replacing the expression of $F(t)$ in the equation of motion, we obtain $m\ddot{x} + kx + \alpha\dot{x} = 0$, which we can put in the form

$\ddot{x} + 2\sigma\dot{x} + \omega_0^2 x = 0$, (with $2\sigma = \dfrac{\alpha}{m}$ and $\omega_0^2 = \dfrac{k}{m}$).

This equation constitutes the equation of motion of damped linear oscillations.

We seek the solution of the equation

From the differential equation $\ddot{x} + 2\sigma\dot{x} + \omega_0^2 x = 0$

The characteristic equation is $r^2 + 2\sigma.r + \omega_0^2 = 0$

(It has the shape of $ar^2 + b.r + c = 0$), its discriminate is $\Delta = \sigma^2 - \omega_0^2$.

Its roots are according to the sign of Δ.

i) 1st case $\Delta < 0$ (which corresponds to the case where $\sigma \prec \omega_0$)

The characteristic equation admits two conjugate complex roots,

$r_1 = -\sigma + j\sqrt{\omega_0^2 - \sigma^2}$ and $r_2 = -\sigma - j\sqrt{\omega_0^2 - \sigma^2}$.

The solution of the differential equation is

$$x(t) = A.e^{r_1 t} + B.e^{r_2 t} = e^{-\sigma t}[A.e^{j\sqrt{\omega_0^2 - \sigma^2}} + B.e^{-j\sqrt{\omega_0^2 - \sigma^2}}].$$

Note 2

The imaginary parts of the complex exponentials give a sinusoid and the real parts of the exponentials give a damping envelope. We can also write the solution in the form $x(t) = a.\cos(\Omega t + \varphi).e^{-\sigma t}$.

a is the amplitude, $\Omega = \sqrt{\omega_0^2 - \sigma^2}$ is the pseudo-pulsation and φ is the phase that depends on the initial conditions (IC).

The movement is damped oscillatory "pseudo-periodic regime".

The corresponding pseudo-period is $T = \dfrac{2\pi}{\Omega} = \dfrac{2\pi}{\sqrt{\omega_0^2 - \sigma^2}}$,

$$T = \frac{2\pi}{\omega_0\sqrt{1-\sigma^2/\omega_0^2}} = \frac{T_0}{\sqrt{1-\sigma^2/\omega_0^2}},$$

$T > T_0$ (own period of the system).

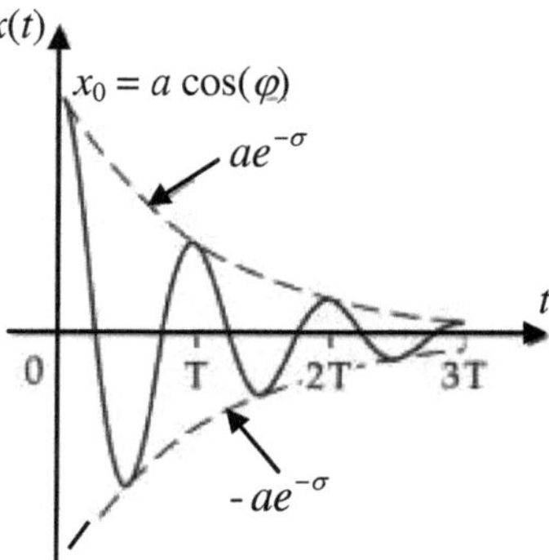

Logarithmic Decrement (LD)

We compare $x(t+T)$ to $x(t)$

$$\frac{x(t+T)}{x(t)} = \frac{a.\cos(\Omega(t+T)+\varphi).e^{-\sigma T}.e^{-\sigma t}}{a.\cos(\Omega t+\varphi).e^{-\sigma t}} = e^{-\sigma T},$$

By definition, the logarithmic decrement is the quantity

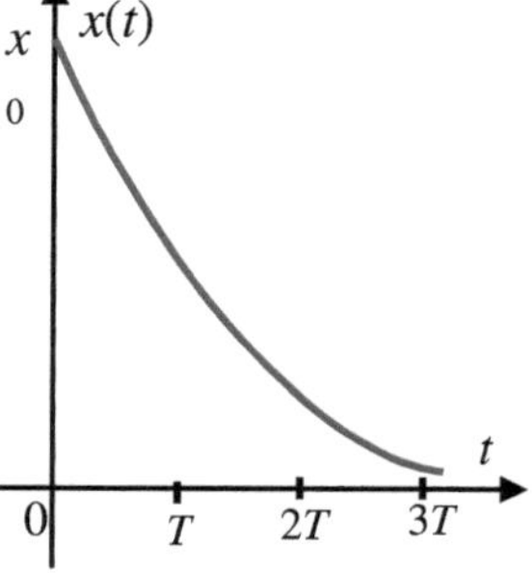

$$LD = \delta = -\ln[\frac{x(t+T)}{x(t)}] = \sigma T.$$

It characterizes the decrease in maximum elongations at each period.

ii) 2nd case $\Delta > 0$ (case where $\sigma \succ \omega_0$)

The characteristic equation admits two roots real negative

$$r_1 = -\sigma + \sqrt{\sigma^2 - \omega_0^2} \quad \text{and} \quad r_2 = -\sigma - \sqrt{\sigma^2 - \omega_0^2}.$$

The solution of the differential equation is then

$$x(t) = A.e^{r_1 t} + B.e^{r_2 t} = e^{-\sigma t}[A.e^{\sqrt{\sigma^2-\omega_0^2}} + B.e^{-\sqrt{\sigma^2-\omega_0^2}}] \quad \text{(both exponentials tend}$$

to zero).

Conclusion

In this case the damping is very important, therefore there are no oscillations; the movement $x(t)$ is called aperiodic regime.

iii) 3rd case $\Delta = 0$, (that's to say $\sigma = \omega_0$)

In this case, the characteristic equation admits a double real root $r_1 = r_2 = \dfrac{-b}{a} = -\sigma$, the solution is of the form $x(t) = (At + B).e^{-\sigma t}$,

(*A* and *B* are constants dependent on the initial conditions).

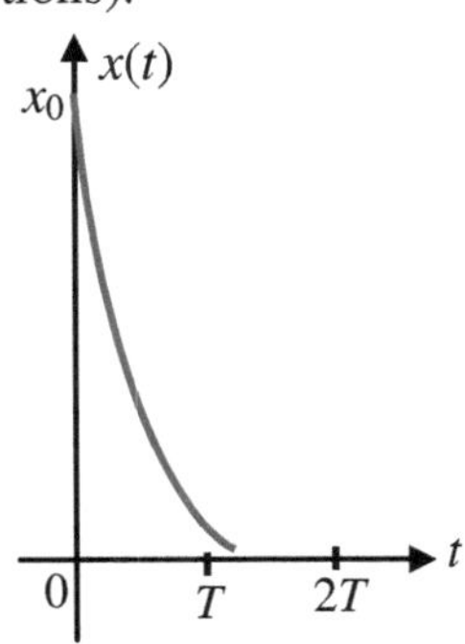

Conclusion
In this 3rd case, there are no oscillations
and the movement is said to be critical.
This corresponds to a borderline case between
the two pseudo-periodic and aperiodic regimes
previously mentioned.

Note 3
The constant $\dfrac{1}{\sigma} \equiv \tau$ is called the relaxation time of the amplitude.

In electricity, $\dfrac{1}{\sigma}$ is called the relaxation time of the circuit.

4.3. Forced oscillations
We consider the general case of a damped and forced oscillation. The Lagrangian is written $L = \dfrac{1}{2}m\dot{x}^2 - \dfrac{1}{2}kx^2 + xF(t)$.

$F(t)$ is a force not deriving from a potential, so the Lagrange equations will take the following form: $\dfrac{d}{dt}\left(\dfrac{\partial L}{\partial \dot{x}}\right) - \dfrac{\partial L}{\partial x} = \sum generalized\ forces\ F(t)$.

This gives the following equation of motion $m\ddot{x} + kx = F(t)$.

We assume that the force $F(t)$ breaks down into two parts $F(t) = F_1(t) + F_2(t)$,

$F_1(t)$ force due to damping $F_1 = -\alpha\dot{x} = -\alpha\dfrac{dx}{dt}$,

$F_2(t)$ force due to permanent external excitation.

4.3.1. Case of sinusoidal excitation
In this case the force is written as $F_2(t) = F_0\cos(\omega t + \varphi) \equiv F_0 e^{j(\omega t + \varphi)}$, then the equation of motion will take the following form
$$m\ddot{x} + kx + \alpha\dot{x} = F_0\cos(\omega t + \varphi).$$

We write it $\ddot{x} + \omega_0 x + 2\sigma \dot{x} = \Gamma \cos(\omega t + \varphi) = \Gamma e^{j(\omega t + \varphi)}$,

(with $\omega_0^2 = \dfrac{k}{m}$, $2\sigma = \dfrac{\alpha}{m}$ and $\Gamma = \dfrac{F_0}{m}$).

The solution of the differential equation is obtained by the superposition of the solution of the homogeneous equation (without second member) and of the particular solution of the equation (with second member).

    *i) **Solution without second member*** (transient response)

$$x(t) = \begin{cases} A\cos(\Omega t + \varphi).e^{-\sigma t} & pseudo-periodic\ regime \\ (At + B).e^{-\sigma t} & aperiodic\ critical\ regime\ . \\ A.e^{r_1 t} + B.e^{r_2 t} & aperiodic\ regime \end{cases}$$

(These solutions are discussed in the previous paragraph).

    *ii) **Solution with second member*** (permanent regime)
We choose a solution of the same form as the second member
$x(t) = \mathrm{Re}(Z) = \mathrm{Re}\left\{ \tilde{Z}_0 .e^{j(\omega t + \varphi)} \right\}.$

$\Rightarrow x(t)$ is solution, so it verifies the equation of motion
$-\omega^2 \tilde{Z}_0 + 2\sigma j \omega \tilde{Z}_0 + \omega_0^2 \tilde{Z}_0 = \Gamma$ (the notation $\tilde{Z}_0$ is complex amplitude).

Find the expression of the vibration amplitude $\tilde{Z}_0$

$$\tilde{Z}_0 = \frac{\Gamma}{\omega_0^2 - \omega^2 + j(2\sigma\omega)} = \frac{\Gamma[(\omega_0^2 - \omega^2) - j(2\sigma\omega)]}{(\omega_0^2 - \omega^2)^2 + 4\sigma^2\omega^2} \equiv \left|\tilde{Z}_0\right|.e^{j\beta} \text{ (module} \times \text{argument)}.$$

$$\left|\tilde{Z}_0\right| = \frac{\Gamma}{\sqrt{(\omega_0^2 - \omega^2)^2 + 4\sigma^2\omega^2}} \text{ and } tg(\beta) = \frac{2\sigma\omega}{\omega_0^2 - \omega^2} .$$

(We find that the amplitude of the oscillations is proportional to the amplitude of the exciting force and depends on its pulsation ω).

4.3.2. Study of the amplitude $\left|\tilde{Z}_0\right|$ as a function of the pulsation ω

a) **If $\alpha = 0$** $\Rightarrow 2\sigma = \dfrac{\alpha}{m} = 0$ (no frictions).

$$\left|\tilde{Z}_0\right| = \frac{\Gamma}{\omega_0^2 - \omega^2} \rightarrow \frac{d\left|\tilde{Z}_0\right|}{d\omega} = \frac{2\omega\Gamma}{[\omega_0^2 - \omega^2]^2} .$$

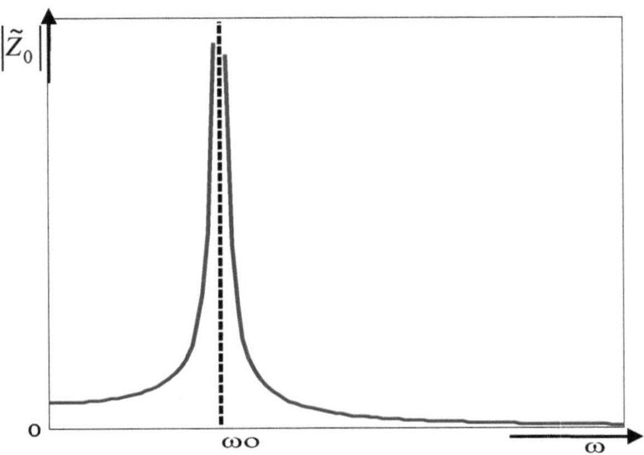

b) If $\alpha \neq 0 \Rightarrow \sigma \neq 0$

$$|\tilde{Z}_0| = \frac{\Gamma}{\sqrt{(\omega_0^2 - \omega^2)^2 + 4\sigma^2\omega^2}} \Rightarrow \frac{d|\tilde{Z}_0|}{d\omega} = \frac{2\sigma\,\Gamma.[(\omega_0^2 - \omega^2)^2 + 2\sigma^2]}{[(\omega_0^2 - \omega^2)^2 + 4\sigma^2\omega^2]^{3/2}}$$

if $[(\omega_0^2 - \omega^2)^2 + 2\sigma^2] = 0 \Leftrightarrow \dfrac{d|\tilde{Z}_0|}{d\omega} = 0$.

So, $\omega = \sqrt{\omega_0^2 - 2\sigma} \equiv \omega_R$ (named *resonance pulsation*).

The resonance pulsation (ω_R) is the pulsation of the exciting force for which the amplitude of the oscillations becomes maximum (It is the value ω of exciter which corresponds to the minimum of the denominator of the amplitude $|\tilde{Z}_0|$).

Values and limits

- Amplitude at resonance corresponds to $\left|\tilde{Z}_0\,(\omega = \omega_R)\right|$.

- Critical case, $\omega_0^2 - 2\sigma^2 = 0 \Rightarrow \sigma = \dfrac{\omega_0}{\sqrt{2}}$, and $|\tilde{Z}_0| = \dfrac{\Gamma}{\sqrt{4\sigma^2\omega^2 - 4\sigma^4}}$.

- The resonance pulsation is determined when $\dfrac{d|\tilde{Z}_0|}{d\omega} = 0$.

$\Rightarrow$ This means that the function $\left|\tilde{Z}_0\,(\omega)\right|$ has a maximum, for $\omega = \omega_R$.

- Moreover, when $\omega \to \infty$. the function $\left|\tilde{Z}_0(\omega)\right| = 0$,

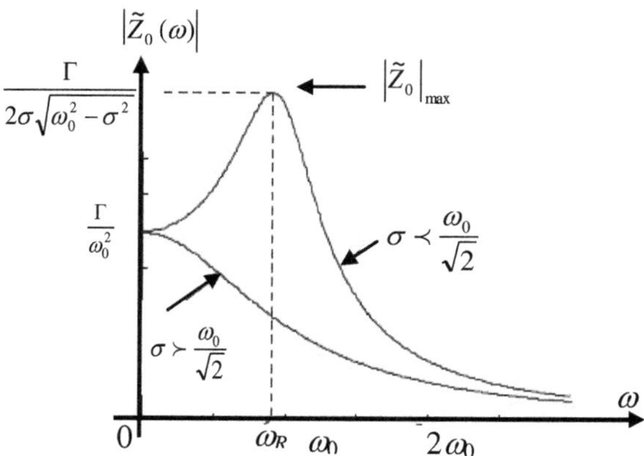

(The plot of $\left|\tilde{Z}_0(\omega)\right|$ as a function of ω).

4.3.3. Effect of damping on vibration

The degree of damping of a system, denoted D, is defined by the ratio $D = \dfrac{\sigma}{\omega_0} = \dfrac{\alpha}{2m\omega_0}$. It is reported that the resonance frequency can cause the rupture (breakage) of a mechanical system if the vibration amplitude is infinite. To remedy this, it is necessary to introduce shock absorbers throughout the vibration system $\alpha \neq 0$ (so the existence of $D \neq 0$).

Therefore, as soon as $D \neq 0$, the maximum vibration amplitude decreases.

To visualize what is happening at the amplitude level, at resonance, a representation of normalized amplitude versus normalized frequency is given below.

We found that the amplitude: $\left|\tilde{Z}_0(\omega)\right| = \dfrac{\Gamma}{\sqrt{(\omega_0^2 - \omega^2)^2 + 4\sigma^2\omega^2}}$,

If the exciting frequency is zero, we get $\left|\tilde{Z}_0(\omega = 0)\right| = \dfrac{\Gamma}{\omega_0^2} = \dfrac{F_0}{k}$.

The representation of the amplitude ratio $\dfrac{\left|\tilde{Z}_0(\omega)\right|}{\left|\tilde{Z}_0(0)\right|}$ as a function of the

pulsation ratio (ω/ω_0) is illustrated, according some values of σ (indicated in legend).

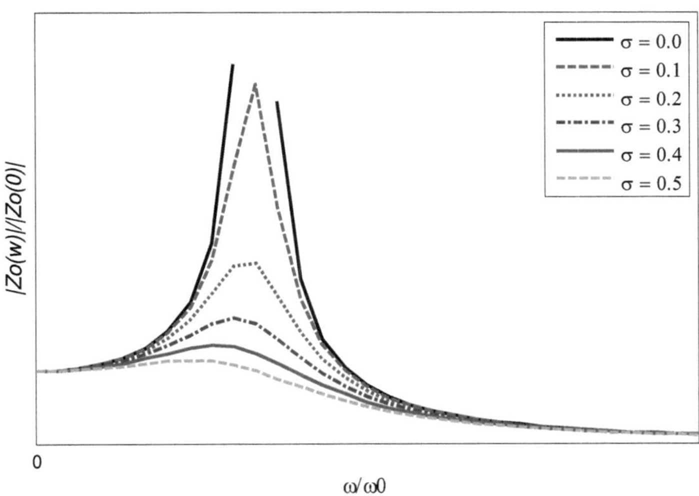

a) Impact of damping on amplitude

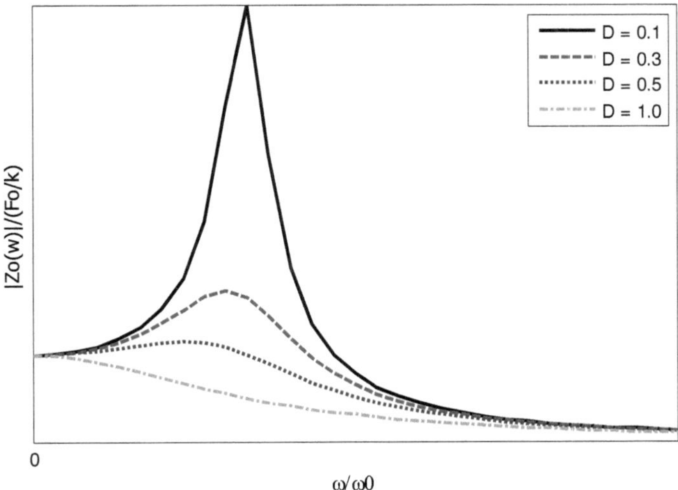

According to the figure, the effect of the damper on the maximum vibration amplitude is noticeable when D increases.

b) Impact of damping on dephasage

The dephasage φ means phase difference between excitation and resulting motion.

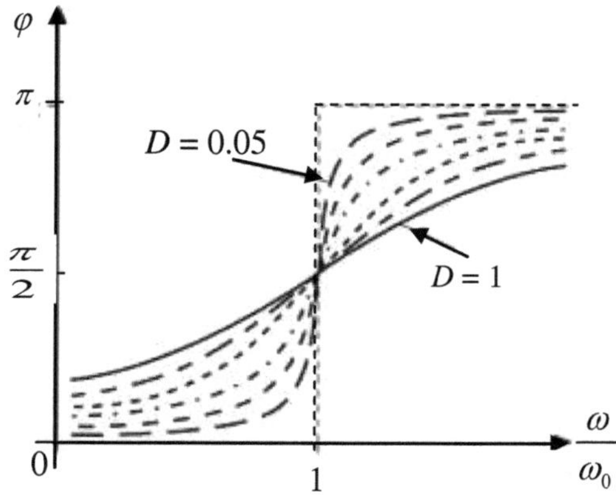

The figure shows the variation of the dephasage as a function of the normalized pulsation $\dfrac{\omega}{\omega_0}$, for different values of damping D.

We observe that for $\omega_{exct} = \omega_0 \Rightarrow$ the dephasage $\varphi = \dfrac{\pi}{2}$ (this property is used to know the value of the pulsation or resonance frequency).

4.4. Resonance of a forced vibration
4.4.1. Obtaining resonance

We observe the phenomenon of resonance, when the exciting pulsation (frequency) tends towards the own pulsation ($\omega \rightarrow \omega_0$) and this happens when the friction is very low ($D \approx 0$). In this case, the amplitude of oscillation tends towards infinity. (The pulse ω is that of the exciter).

Note 4

1) The position of the amplitude resonant frequency (ω_R) shifts toward lower frequencies as the intensity of the damper increases (see the figure illustrating the effect of the damper on the range of motion).

2) In practice, it is necessary to avoid the operation of electrical measuring devices at their resonance frequencies (in the oscillating circuits). Because, said frequency can destroy them and constitutes a danger.

4.4.2. Parameters of a resonance

Each resonance is characterized, mainly, by two parameters:

i) An overvoltage, expressed by $\dfrac{|Z_0(\omega_R)|}{F_0/k} = \dfrac{|Z_{0\,max}|}{\Gamma}$.

ii) The width at half height, it is defined by $\dfrac{|Z_{0\,max}|}{\sqrt{2}}$.

These two parameters
are illustrated on the picture.

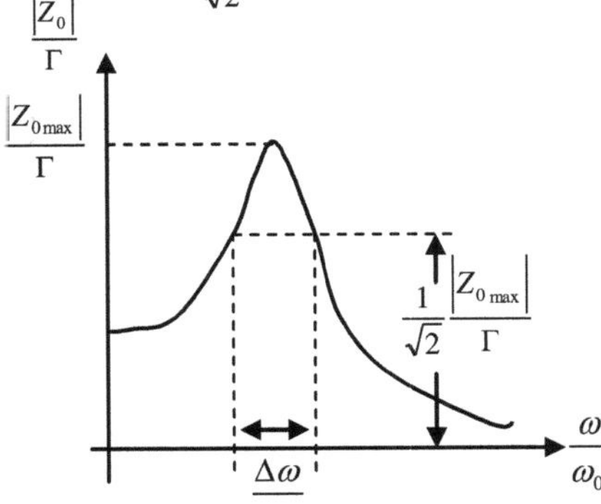

Graphic representation of width at half height $\dfrac{\Delta\omega}{\omega_0}$ and overvoltage $\dfrac{|Z_{0\,max}|}{\Gamma}$ at the resonance, in the case of a forced oscillation.

4.4.3. Bandwidth

The bandwidth at 3 (dB) is, by definition, the set of pulsations for which the amplitude of the movement is $A \geq \dfrac{|Z_{0\,max}|}{\sqrt{2}}$.

The width of the band is $\Delta\omega = \omega_2 - \omega_1$.

Here ω_1 and ω_2 are, respectively, the pulsations for which

$$A(\omega_1) = A(\omega_2) = \dfrac{A_{max}}{\sqrt{2}}.$$

Note 5

In the oscillating electric circuits, we seek to have a strong resonance with a weak damping (see the curve of the amplitude $A(\omega)$).

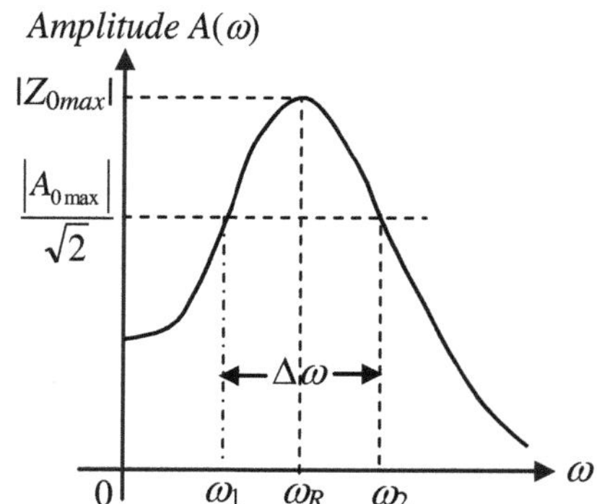

In this case, $\sigma^2 \prec\prec \omega_0$

$\Rightarrow A_{max} = \dfrac{\Gamma}{2\sigma\omega_0}$.

The pulsations ω_1 and ω_2 are defined by

$$\dfrac{|Z_{0max}|}{\sqrt{2}} = \dfrac{\Gamma}{2\sqrt{2}\sigma\omega_0} = A(\omega) = \dfrac{\Gamma}{\sqrt{4\sigma^2\omega^2 + (\omega^2 - \omega_0^2)^2}}$$

$$\Rightarrow 4\sigma^2\omega^2 + (\omega^2 - \omega_0^2)^2 = 8\sigma^2\omega_0^2.$$

Therefore, if σ is small, causes a narrow resonance peak ($\Delta\omega$ small), then we can consider that $\Delta\omega$ and ω close to ω_0.

$$\Rightarrow (\omega^2 - \omega_0^2)^2 = 4\sigma^2\omega_0^2 \text{ and so } \omega^2 - \omega_0^2 = \pm 2\sigma\omega_0.$$

$\Rightarrow$ the two roots $\omega_1^2 = \omega_0^2(1 - 2\dfrac{\sigma}{\omega_0})$ and $\omega_2^2 = \omega_0^2(1 + 2\dfrac{\sigma}{\omega_0})$,

which gives approximately $\omega_1 \approx \omega_0(1 - \dfrac{\sigma}{\omega_0})$ and $\omega_2 \approx \omega_0(1 + \dfrac{\sigma}{\omega_0})$.

Hence, the bandwidth of the oscillator $\Delta\omega = \omega_2 - \omega_1 = 2\sigma = \dfrac{\alpha}{m}$.

4.4.4. Notion of quality coefficient

The quality factor (Q) of a forced oscillator defines the fineness and the amplitude of the peak. It is given by $Q = \dfrac{\omega_0}{\Delta\omega} = \dfrac{\omega_0}{2\sigma}$.

This coefficient characterizes the amount of resonance of the system.

5. Nonlinear oscillations

In this case the function $U(x)$ is not parabolic, but keeps an equilibrium position x_0, where $U(x_0)$ is min. in the case of an oscillation
      - linear if $x_1 x_0 = x_2 x_0$,

- non linear if $x_1 x_0 \neq x_2 x_0$, (x_1 and x_2 are not symmetrical).

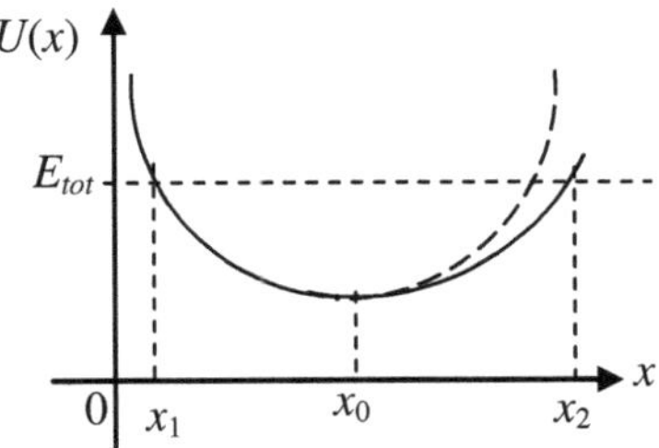

For low energies (small oscillations)

$$U(x) = U(x_0) + \frac{(x-x_0)}{1!}\frac{dU(x)}{dx}\bigg]_{x=x_0} + \frac{(x-x_0)^2}{2!}\frac{d^2U(x)}{dx^2}\bigg]_{x=x_0} +$$

$$\ldots + \frac{(x-x_0)^n}{n!}\frac{d^nU(x)}{dx^n}\bigg]_{x=x_0}$$

Application: case where $n = 4$ and $x_0 = 0$,

$$U(x) = U(0) + x\frac{dU(x)}{dx}\bigg]_{x=0} + \frac{x}{2}\frac{d^2U(x)}{dx^2}\bigg]_{x=0} + \frac{x^3}{6}\frac{d^3U(x)}{dx^3}\bigg]_{x=0} + \frac{x^4}{24}\frac{d^4U(x)}{dx^4}\bigg]_{x=0}$$

$$\Rightarrow \quad L = \frac{1}{2}m\dot{x}^2 - \frac{m\omega_0^2}{2}x^2 - \frac{ma}{3}x^3 - \frac{mb}{4}x^4, \quad a \quad \text{and} \quad b \quad \text{are} \quad \text{constants}$$

homogenous to the pulsations.

(U is a polynomial of degree 4, which we write

$U(x) = A + Bx + Cx^2 + Dx^3 + Ex^4$).

In this type of oscillations (ONL), the period is not a constant but it is expressed as a function of the oscillation amplitude.

6. Oscillating circuits and electromechanical analogy
6.1. Introduction and definitions

The electric circuit is a basic element in electronics. Its association with an antenna makes it possible to produce electromagnetic radiation. The oscillating circuit is obtained by association (in series or in parallel) of an inductance coil L and a capacitor of capacitance C. We can consider that the variable physical quantity is the charge q of the capacitor.

The variation of q as a function of time t produces an electric current that induces a magnetic field in the inductance.

The electrical circuit being closed on itself, the sum of the potential differences is zero. The energy of the circuit is alternately stored in capacitor C and in inductance L.

Therefore, electrical oscillations are described by the same type of 2nd order differential equation as mechanical oscillations.

We can then make analogies between mechanical and electrical quantities.

Mechanical quantity (mass + spring)	Equivalent electrical quantity
Elongation x	Electric charge q
Velocity $v = \dot{x} = \dfrac{dx}{dt}$	Electric current $i = \dot{q} = \dfrac{dq}{dt}$
Acceleration $a = \ddot{x} = \dfrac{d^2x}{dt^2}$	2nd derived $\ddot{q} = \dfrac{d^2q}{dt^2}$
Mass m	Self (coil) L
Stiffness k	Inverse of capacitance $\dfrac{1}{C}$
Shock absorber α	Resistance R
Own period $T_p = 2\pi\sqrt{\dfrac{m}{k}}$	Own period $T_p = 2\pi\sqrt{LC}$
Kinetic energy $E_c = \dfrac{1}{2}m\dot{x}^2$	Coil energy $E_L = \dfrac{1}{2}L\dot{q}^2$
Elastic energy $E_k = \dfrac{1}{2}kx^2$	Capacitor energy $E_C = \dfrac{1}{2C}q^2$
Exciting force $F(t)$	Power voltage $U(t)$

Example

The study of the oscillating L-C electrical circuit is done by applying Kirchhoff's laws.

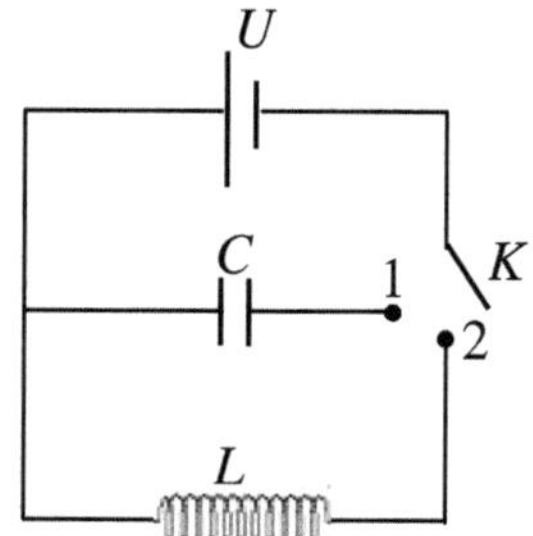

Switch on the position 1

The condenser C, subjected to the tension U of the generator, is charged.

If we denote the electric charge by q, the voltage across C is U_C, such that

$$U_C = U = \frac{q}{C}.$$

Switch on the position 2

The capacitor C discharges into inductor L, and a current $i = \dfrac{dq}{dt}$ circulates in the mesh.

Along the closed mesh $U_C + U_L = 0 \Rightarrow \dfrac{q}{C} + L\dfrac{di}{dt} = 0$,

On the other hand, $i = \dfrac{dq}{dt}$, then $\dfrac{q}{C} + L\dfrac{d^2q}{dt^2} = 0$.

This gives a homogeneous 2nd order differential equation

$$\ddot{q} + \frac{1}{LC}q = 0, \text{ (like mechanical oscillations, we pose } \omega_0^2 = \frac{1}{LC}).$$

6.2. Use of electric oscillators

When the mechanical systems present a great complexity, the equivalent electrical system makes it possible to overcome the difficulty. At the end, we will deduce the characteristic quantities of the system.

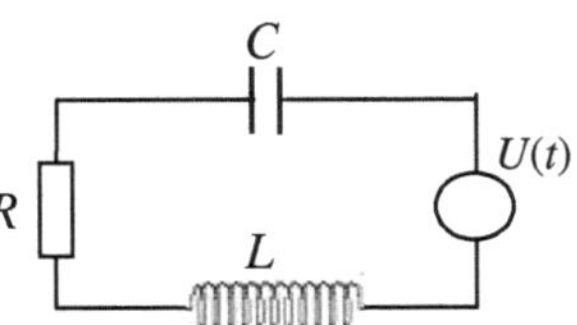

Example of application

We give the following forced and damped mechanical system

$$\vec{F}(t) = F_0 \cos(\omega t)\vec{i}$$

$$\vec{f} = -\alpha\, \dot{x}\vec{i}$$

The electrical circuit equivalent to the mechanical assembly is

The voltage across L is $U_L = L\dfrac{di}{dt}$.

Resistance dissipates energy (Joule effect) $U_R = Ri$.

The voltage across C is $U_C = \dfrac{q}{C} = \dfrac{1}{C}\int i\,dt$.

Applying Ohm's law to the mesh gives

$$U_C + U_L + U_R = U(t) \Rightarrow \frac{1}{C}\int i\,dt + L\frac{di}{dt} + Ri = U(t).$$

We write the equation in terms of q, $\quad L\frac{d^2q}{dt^2} + R\frac{dq}{dt} + \frac{1}{C}q = U_0\cos(\omega t).$

The equation can also be written in the classical form

$$\ddot{q} + 2\sigma\dot{q} + \omega_0^2 q = \Gamma_0\cos(\omega t), \text{ with } 2\sigma = \frac{R}{L}, \ \omega_0^2 = \frac{1}{LC} \text{ and } \Gamma_0 = \frac{U_0}{L}.$$

- the degree of damping of the circuit is $D = \dfrac{\sigma}{\omega_0} = \dfrac{R}{2}\sqrt{\dfrac{C}{L}}$,

- circuit quality factor is $\Gamma_0 = \dfrac{1}{R}\sqrt{\dfrac{L}{C}}$.

Note 6

1) From the results of the electrical oscillations, we can go back to the dynamics of the mechanical oscillator, based on the equivalences between elements.

From the charge equation $L\dfrac{d^2q}{dt^2} + R\dfrac{dq}{dt} + \dfrac{1}{C}q = U_0\cos(\omega t).$

The equation of the mechanical system is $m\dfrac{d^2x}{dt^2} + \alpha\dfrac{dx}{dt} + kx = F_0\cos(\omega t).$

(Because, $L \leftrightarrow m$, $R \leftrightarrow \alpha$, $1/C \leftrightarrow k$, $q(t) \leftrightarrow x(t)$ et $U(t) \leftrightarrow F(t)$).

2) It is also possible to determine the characteristic quantities of the oscillation of the system

- the damping coefficient $2\sigma = \dfrac{\alpha}{m}$,

- the own pulsation $\omega_0^2 = \dfrac{k}{m}$ and the constant $\Gamma_0 = \dfrac{F_0}{m}$.

Chapter 4

Oscillations of systems with two degrees of freedom

1. Oscillators with two degrees of freedom (2 dof)
1.1. Reminders
An oscillator (at one dof) has a single own frequency of vibration determined by its characteristics. In oscillators with two dof or more (case of coupled oscillators for example), they have several ways of oscillating called eigenmodes.

1) During the study of the vibrational dynamics of systems with one degree of freedom, we found the existence of *a single eigenmode of vibration* ω_0.

(Pulsation expressed as $\omega_0 = \sqrt{\dfrac{k}{m}}$, $\omega_0 = \sqrt{\dfrac{g}{l}}$, ...).

2) For a system with two (2) degrees of freedom $\Rightarrow$ we should have two (2) eigenmodes of vibration.

1.2. Application example
Consider two mass-spring type systems coupled by a third spring.

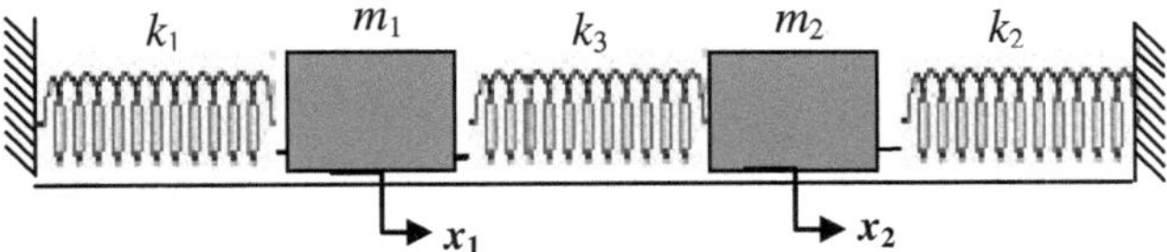

Figure 1: Coupling of two harmonic oscillators by a spring (elastic coupling).

In this case, we have the following energies

$$\begin{cases} T = \dfrac{1}{2}m_1\dot{x}_1 + \dfrac{1}{2}m_2\dot{x}_2 \\ U = \dfrac{1}{2}k_1 x_1^2 + \dfrac{1}{2}k_2 x_2^2 + \dfrac{1}{2}k_3(x_2 - x_1)^2 \end{cases}.$$

The Lagrange equations (of motion) are then given by

$$\begin{cases} \dfrac{d}{dt}\left(\dfrac{\partial L}{\partial \dot{x}_1}\right) - \dfrac{\partial L}{\partial x_1} = 0 \\[2mm] \dfrac{d}{dt}\left(\dfrac{\partial L}{\partial \dot{x}_2}\right) - \dfrac{\partial L}{\partial x_2} = 0 \end{cases} \Rightarrow \begin{cases} m_1 \ddot{x}_1 + (k_1 + k_3)x_1 = k_3 x_2 \\[2mm] m_2 \ddot{x}_2 + (k_2 + k_3)x_2 = k_3 x_1 \end{cases}.$$

They represent a system of coupled differential equations.

They are written in the form $\begin{cases} \ddot{x}_1 + \omega_1^2 x_1 = \omega_{01}^2 x_2 \\ \ddot{x}_2 + \omega_2^2 x_2 = \omega_{02}^2 x_1 \end{cases}$, (with $\omega_1^2 = \dfrac{k_1 + k_3}{m_1}$,

$\omega_{01}^2 = \dfrac{k_3}{m_1}$ $\omega_2^2 = \dfrac{k_2 + k_3}{m_2}$, $\omega_{02}^2 = \dfrac{k_3}{m_2}$).

Switch to matrix writing $\begin{pmatrix} \ddot{x}_1 \\ \ddot{x}_2 \end{pmatrix} = \begin{bmatrix} -\omega_1^2 & \omega_{01}^2 \\ \omega_{02}^2 & -\omega_2^2 \end{bmatrix}\begin{pmatrix} x_1 \\ x_2 \end{pmatrix}$, then we look for

solutions in the form $x_1(t) = A_1 . e^{j\Omega t}$ and $x_2(t) = A_2 . e^{j\Omega t}$.

This will then give $\begin{cases} -\Omega^2 A_1 + \omega_1^2 A_1 = \omega_{01}^2 A_2 \\ -\Omega^2 A_2 + \omega_2^2 A_2 = \omega_{02}^2 A_1 \end{cases}$

$$\Rightarrow \begin{bmatrix} -\Omega^2 + \omega_1^2 & -\omega_{01}^2 \\ -\omega_{02}^2 & -\Omega^2 + \omega_2^2 \end{bmatrix}\begin{bmatrix} A_1 \\ A_2 \end{bmatrix} = \begin{bmatrix} 0 \\ 0 \end{bmatrix}.$$

The system admits non-trivial solutions such as
$A_1 = A_2 = 0$ (no physical meaning), determinant of the system must be zero.

$$\det\begin{bmatrix} -\Omega^2 + \omega_1^2 & -\omega_{01}^2 \\ -\omega_{02}^2 & -\Omega^2 + \omega_2^2 \end{bmatrix} = 0$$

$$\Leftrightarrow (-\Omega^2 + \omega_1^2)(-\Omega^2 + \omega_2^2) - (-\omega_{01}^2)(-\omega_{02}^2) = 0$$

$$\Omega^4 - \Omega^2(\omega_1^2 + \omega_2^2) + \omega_1^2.\omega_2^2 - \omega_{01}^2.\omega_{02}^2 = 0,$$

$$\Delta = (\omega_1^2 + \omega_2^2)^2 - 4.(\omega_1^2.\omega_2^2 - \omega_{01}^2.\omega_{02}^2) = 0.$$

We obtain the two (2) roots Ω_1 and Ω_2 which are the eigen-pulsations of the coupled system (the eigenvalues of the matrix).

$$\Omega_1 = \frac{(\omega_1^2 + \omega_2^2) + \sqrt{\Delta}}{2} \quad \text{and} \quad \Omega_2 = \frac{(\omega_1^2 + \omega_2^2) - \sqrt{\Delta}}{2}$$ (the system can oscillate

with Ω_1, Ω_2 or a linear combination of the two modes.

Each solution Ω_i describes a particular of the motion.

The general solution is written as $x_i(t) = \sum_i \tilde{A}_1 . e^{j\Omega it}$.

$$\begin{cases} x_1(t) = A_{11} e^{j\Omega_1 t} + A_{12} e^{j\Omega_2 t} \\ x_2(t) = A_{21} e^{j\Omega_1 t} + A_{22} e^{j\Omega_2 t} \end{cases} \equiv \begin{cases} x_1(t) = A_{11} \cos(\Omega_1 + \varphi_1) + A_{12} \cos(\Omega_2 + \varphi_2) \\ x_2(t) = A_{21} \cos(\Omega_1 + \varphi_1) + A_{22} \cos(\Omega_2 + \varphi_2) \end{cases}$$

Discussions

Case where $m_1 = m_2$ et $k_1 = k_2$ (the coupling of the two oscillators is identical).

$$\begin{cases} \omega_1 = \omega_2 = \omega_0' \\ \omega_{01} = \omega_{02} = \omega_0 \end{cases} \rightarrow \Delta = 4\omega_0'^4 - 4(\omega_0'^4 - \omega_0^4) = 4\omega_0^4$$

The two own pulsations are $\Omega_1 = \omega_0'^2 + \omega_0^2$ and $\Omega_2 = \omega_0'^2 - \omega_0^2$.

- The two oscillators vibrate with the 1st mode Ω_1

$$\begin{cases} x_1(t) = A_{11} \cos(\Omega_1 t + \varphi_1) \\ x_2(t) = A_{21} \cos(\Omega_1 t + \varphi_1) \end{cases} ;$$

Substituting in the equations of motion, we get

$$\begin{cases} \ddot{x}_1 + \omega_0'^2 x_1 = \omega_0^2 x_2 \\ \ddot{x}_2 + \omega_0'^2 x_2 = \omega_0^2 x_1 \end{cases} \Rightarrow \Omega_1^2 A_{11} + \omega_c'^2 A_{11} = \omega_0^2 A_{21} = A_{11}(\omega_0'^2 - \Omega_1^2),$$

$$\Rightarrow A_{21} = A_{11} \frac{\omega_0'^2 - \Omega_1^2}{\omega_0^2} = A_{11} \frac{\omega_0'^2 - \omega_0'^2 - \omega_0^2}{\omega_0^2} = -A_{11}.$$

Then, $x_1(t) = -x_2(t) \Leftrightarrow x_1(t) = e^{j\pi}.x_2(t)$ (vibration in phase opposition, i.e.: when x_1 is max, x_2 is min and vice versa).

The displacements are then written as follows

$$\begin{cases} x_1(t) = A_{11} \cos(\Omega_1 t + \varphi_1) \\ x_2(t) = A_{11} \cos(\Omega_1 t + \varphi_1 + \pi) \end{cases} .$$

- The frequency vibration is Ω_1 (2nd mode)

$$\begin{cases} x_1(t) = A_{12} \cos(\Omega_2 t + \varphi_2) \\ x_2(t) = A_{22} \cos(\Omega_2 t + \varphi_2) \end{cases} ; \text{(Same steps as before)},$$

We will find $A_{12} = A_{22}$ (in-phase vibration).

The constants A_{11} and A_{12} are determined by the initial conditions.

(For example, $x_1(0) = x_0$, $x_2(0) = 0$ and $\dot{x}_1(0) = \dot{x}_2 = 0$,

Just replace in the equations the time t by 0, then we will calculate A_{12}, A_{22} according to x_0 and ω_0, ...).

Conclusion

The vibration eigenmodes of two coupled oscillators designate the oscillation modes during which the oscillators do not exchange energy.
- if the movements of the two coupled oscillators are constantly parallel.
$\Rightarrow$ We speak of parallel oscillators (or in phase).
- If the movements of the two coupled oscillators are in opposition
$\Rightarrow$ We speak of antiparallel oscillators (or in opposite phases).

2. Coupling notion

Coupled oscillators consist of at least two oscillators influencing each other via a coupling device (which can be a mass, a spring, a damper, etc.). Oscillators formed by coupling exchange their energy.

By definition, the coupling coefficient K is given by $K = \left[\dfrac{\Omega_1^2 - \Omega_2^2}{\Omega_1^2 + \Omega_2^2}\right]^{1/2}$,

where $0 \leq K \leq 1$.

The pulsations Ω_1 et Ω_2 are the own pulsations of the oscillators.

In the example considered, we find that $K = \left[\dfrac{k_3^2}{(k_1 + k_3)(k_2 + k_3)}\right]^{1/2}$.

Discussions

- If $k_3 = 0 \Rightarrow K = 0$, the coupling is zero, the movements of the masses m_1 and m_2 are independent, the coupling is said to be released.

- If $k_3 \gg \Rightarrow K \approx 1$, the spring k_3 is replaced by a rigid bar, the two masses m_1 and m_2 are united and the system constitutes a single oscillator, the coupling is said to be tight (there is one degree of freedom in this case).

3. Superposition of two vibrations

Since the equations of motion of two harmonic oscillators are linear, then we can superimpose them.

We write $x(t) = x_1(t) + x_2(t)$, (we recall that x_1 and x_2 are the solutions of the differential equations of motion).

Three possibilities can be distinguished

3.1. Both oscillators vibrate at the same pulsation

We assume that the movements of the two harmonic oscillators are described by the following functions

$$x_1(t) = A_1 \sin(\omega t + \varphi_1) \text{ and } x_2(t) = A_2 \sin(\omega t + \varphi_2).$$

The addition of the two functions gives the following resultant (which represents a harmonic oscillation also):

$$x(t) = x_1 + x_2 = A_1 \sin(\omega t + \varphi_1) + A_2 \sin(\omega t + \varphi_2) = A \sin(\omega t + \varphi).$$

(The resulting amplitude is: $A = \sqrt{A_1^2 + A_2^2 + 2A_1 A_2 \cos(\varphi_1 - \varphi_2)}$, and the

resulting dephasage is $\varphi = \arctan\left[\dfrac{A_1 \sin(\varphi_1) + A_2 \sin(\varphi_2)}{A_1 \cos(\varphi_1) + A_2 \cos(\varphi_2)}\right]$).

Example

To visualize the function $x(t)$, which constitutes the superposition, we

consider the following functions: $x_1(t) = 2\sin(2t + \dfrac{\pi}{3})$ and $x_2(t) = 3\sin(2t)$,

$\Rightarrow x(t) = x_1 + x_2 = \sqrt{19}\sin(2t + 0.13\pi)$.   In this case, the ratio of the

amplitudes $\dfrac{A_1}{A_2} = \dfrac{2}{3}$, and the dephasage $\varphi = 0.13\pi$. The plot of the three

functions is given in figure 2.

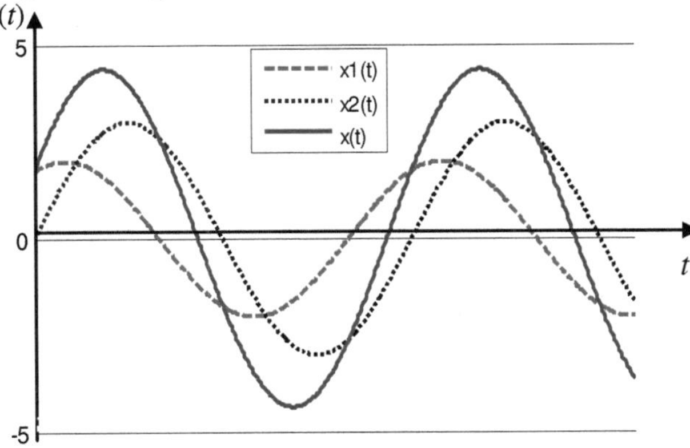

Figure 2: Illustrative plot of an example of superposition of two sinusoids x_1 and x_2, of the same pulsation ω, of the phase shift $\varphi = 0.13\pi$ and

amplitude ratio $\dfrac{A_1}{A_2} = \dfrac{2}{3}$.

Particular cases

Depending on the value of the phase shift φ, we speak of constructive or destructive superimpositions of oscillations.

a) If the two oscillators vibrate in phase with different amplitudes $A_1 \neq A_2$, and the dephasage $\varphi = 0$, we get the following plot

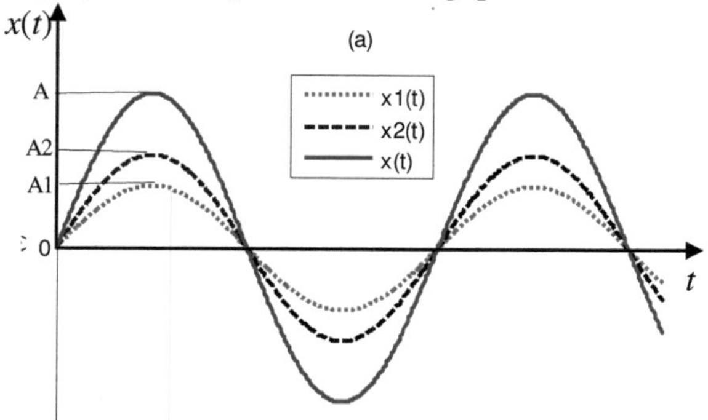

b) If the amplitudes of the two oscillators are identical $(\frac{A_1}{A_2} = 1)$ and the movements of the two oscillators in phase opposition $\varphi = \pi$, we obtain

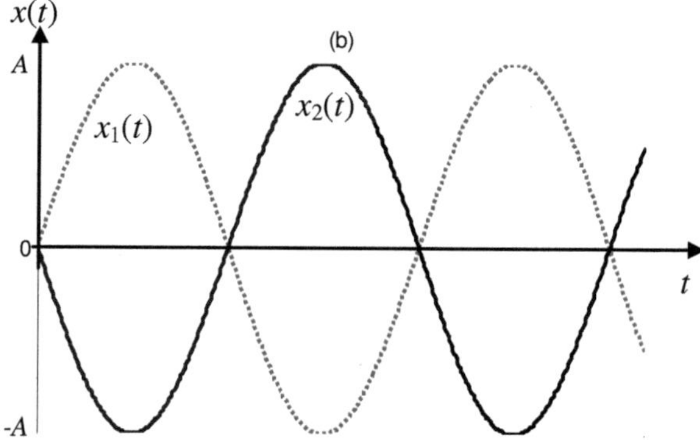

Figure 3: Superposition of two oscillations x_1 and x_2 of the same pulsation.
 (a) Case where the amplitudes are different and $\varphi = 0$
 (The superposition is said to be constructive),
 (b) Case where the amplitudes are identical and $\varphi = \pi$,
 (The superimposition is said to be destructive).

3.2. The oscillators vibrate with different pulsations
a) Addition of two oscillators with different pulsations
To simplify the calculations, we consider the case where the two oscillators are in phase and carry out vibrations of the same amplitude

We write $x_1(t) = A\sin(\omega_1 t)$ and $x_2(t) = A\sin(\omega_2 t)$.

The sum $x(t) = x_1 + x_2 = A\sin(\omega_2 t) + A\sin(\omega_2 t)$,

Which ultimately gives $x(t) = 2A\cos(\dfrac{\omega_1 - \omega_2}{2}t) \times \sin(\dfrac{\omega_1 + \omega_2}{2}t)$.

Generally, the resultant is not a harmonic oscillation.

Note 1
If the pulsations (ω_1 and ω_2) of two superimposed vibrations are very close, the "beat" phenomenon is observed. It appears when the system does not oscillate in one or the other of the eigenmodes of the system.

It manifests as follows:

If one of the harmonic oscillators is moved away from its equilibrium position, then abandoned without initial speed, while the other oscillator is left at rest, we see that the first oscillator will excite the second whose amplitudes will increase. That is to say, the energy of the first oscillator is gradually transferred to the second, and then the process is reversed again and so on.

Interpretation
The result of the superimposition, found earlier, can be decomposed as

$$x(t) = 2A\underbrace{\cos(\dfrac{\omega_1 - \omega_2}{2}t)}_{amplitude} \times \sin(\dfrac{\omega_1 + \omega_2}{2}t).$$

It can be thought of as a pulsating vibration $(\dfrac{\omega_1 + \omega_2}{2})$ and whose amplitude varies periodically with the pulsation $(\dfrac{\omega_1 - \omega_2}{2})$.

b) Beat period
It defines the time between two consecutive zero crossings of the beat amplitude. It is given by $T_s = \dfrac{2\pi}{\omega_1 - \omega_2}$.

Important

Please note that ***beat period*** should not be confused with ***period of beats***.

The beat period $T_s = \dfrac{2\pi}{\omega_1 - \omega_2}$, as for the period of the beats is expressed by

$$T = \frac{2\pi}{\omega} = \frac{2\pi}{\dfrac{\omega_1 + \omega_2}{2}} = 2\frac{T_1 T_2}{T_1 + T_2}\ .$$

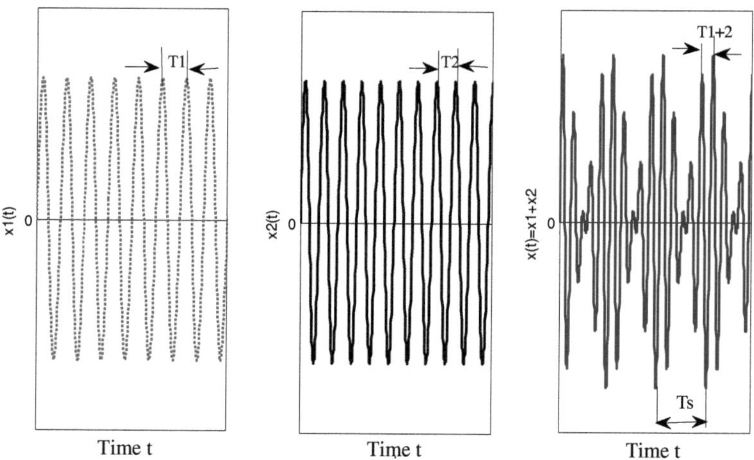

Figure 4: Beating phenomenon generated by the two oscillators $x_1(t)$ and $x_2(t)$. (example $\Delta\omega = \omega_1 - \omega_2 = 2$), for a phase shift $\varphi = 0$ and an amplitude ratio $\dfrac{A_1}{A_2} = 1$.

The taken functions are $x_1(t) = A\sin(10t)$ and $x_2(t) = A\sin(8t)$.

$$\Rightarrow x(t) = x_1 + x_2 = 2A\cos(\frac{10-8}{2}t)\times\sin(\frac{10+8}{2}t).$$

In the figure, the symbol T_s is the period of the beats and T_{1+2} is the period of the resulting oscillation.

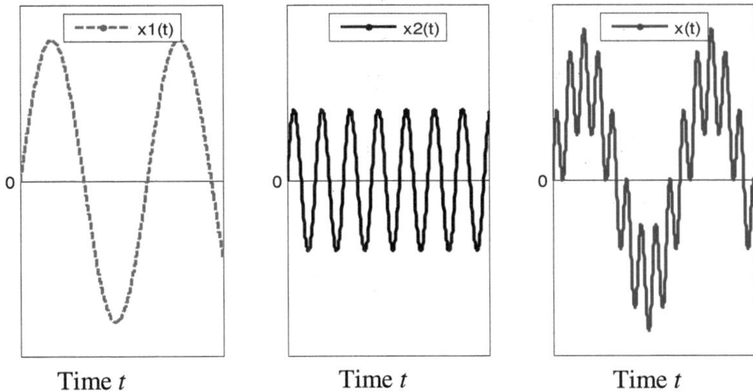

Figure 5: Superposition of two vibrations $x_1(t)$ and $x_2(t)$ with very different pulsations (example $\Delta\omega = \omega_1 - \omega_2 = 9$), for a phase shift $\varphi = 0$ and an amplitude ratio $\dfrac{A_1}{A_2} = 2$.

3.3. Superposition of oscillators of different pulsations and directions

When the two oscillations have different directions (for example, propagation along the x-axis for one and along the y-axis for the other) are superimposed, their independent movements can be easily described

In Cartesian coordinates

$$x(t) = A_x \sin(\omega_x t + \varphi_x) \quad \text{and} \quad y(t) = A_y \sin(\omega_y t + \varphi_y).$$

In polar coordinates

$$r(t) = \sqrt{x^2 + y^2} \quad \text{and} \quad \theta(t) = \arctan\left(\frac{y(t)}{x(t)}\right).$$

The r designates the length of the resulting vector and θ its polar angle along the x-axis.

The extremity of the vector $\vec{r}$ describes a curve called the Lissajous figure, whose design depends on the amplitudes $\dfrac{A_x}{A_y}$, the pulsations $\dfrac{\omega_x}{\omega_y}$ and the phase difference $\Delta\varphi = \varphi_x - \varphi_y$.

Lissajous's figures can be viewed using an oscilloscope with independent horizontal x and vertical y inputs and two sinusoidal voltage generators.

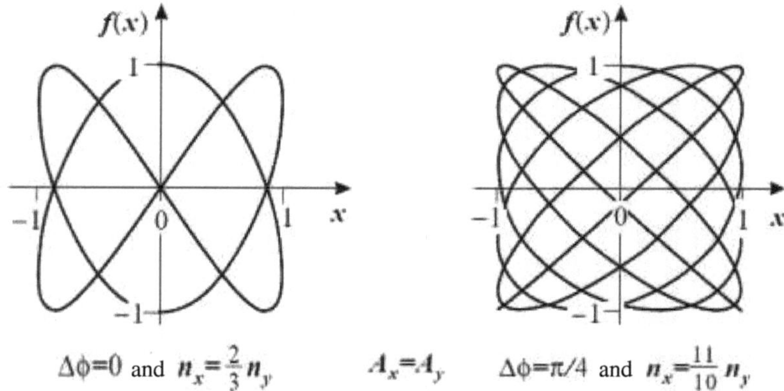

$$\Delta\phi=0 \text{ and } n_x=\frac{2}{3}n_y \qquad A_x=A_y \qquad \Delta\phi=\pi/4 \text{ and } n_x=\frac{11}{10}n_y$$

Figure 6: Lissajous's figures.

4. Oscillations of systems with n degrees of freedom
For a system with two (2) degrees of freedom, we note:
- two (2) own modes of vibration,
- the general solution, was presented in the form of a linear combination of the two (2) modes

Generalization: *case of a system with N degrees of freedom*
We will necessarily obtain N eigenmodes of vibration.

Indeed, the energies are given by
$$\begin{cases} T = \dfrac{1}{2}\sum_{i=1}^{N}\sum_{j=1}^{N} m_{ij}\dot{x}_i\,\dot{x}_j \\ U = \dfrac{1}{2}\sum_{i=1}^{N}\sum_{j=1}^{N} k_{ij}x_i\,x_j \end{cases}.$$

Lagrange's equations give the system of differential equations characterizing the motion of the system
$$\begin{cases} m_{11}\ddot{x}_1 + m_{12}\ddot{x}_2 + ... + m_{1N}\ddot{x}_N + k_{11}x_1 + k_{12}x_2 + ... + k_{1N}x_N = 0 \\ m_{21}\ddot{x}_1 + m_{22}\ddot{x}_2 + ... + m_{2N}\ddot{x}_N + k_{21}x_1 + k_{22}x_2 + ... + k_{2N}x_N = 0 \\ \vdots \\ m_{N1}\ddot{x}_1 + m_{N2}\ddot{x}_2 + ... + m_{NN}\ddot{x}_N + k_{N1}x_1 + k_{N2}x_2 + ... + k_{NN}x_N = 0, \end{cases}$$

Since we are in the case of free linear oscillations, then the solutions must be chosen according to this type $x_i(t) = A_i e^{j\omega t}$.

In detail
$$\begin{cases} x_1(t) = A_1 e^{j\omega t} \\ \vdots \\ x_N(t) = A_N e^{j\omega t} \end{cases}$$
to be replaced in the equations of motion.

We get the following matrix
$$\underbrace{\begin{bmatrix} -\omega^2 + \omega_1 & \cdots & \\ \vdots & \cdots & \vdots \\ & \cdots & -\omega^2 + \omega_N \end{bmatrix}}_{dynamics\ matrix} \begin{bmatrix} A_1 \\ \vdots \\ A_N \end{bmatrix} = \begin{bmatrix} 0 \\ \vdots \\ 0 \end{bmatrix}.$$

The eigenvalues of the dynamic matrix give access to the N eigenmodes of vibration.

Once the eigenfrequencies have been determined, we report each of them in the system of equations; this will make it possible to have the values of the A_i (using the initial conditions).

We proceed in the same way as the case of two (2) degrees of freedom.

Note 2

If the system is forced and/or damped, the expressions of the forces will be introduced then the appropriate Lagrange equations will be used.

Exercise series n° 2 (Vibrations)

Exercise 2.1

Give the generalized coordinates appropriate to the following systems:
1) A material point subject to moving on an ellipse.
2) A circular cylinder rolling on an inclined plane.
3) A simple pendulum of mass m.
4) A double mass pendulum m_1 and m_2.
5) A plane pendulum of mass m_2 and the suspension point m_1 moves on a straight line according to the following cases:

    i) the suspension point moves uniformly on a vertical circle;

    ii) the suspension point performs alternately horizontal and vertical oscillations of the shape $a.\cos(\theta)$.
6) Atwood's machine.

Exercise 2.2

A material point M of mass m is mobile on an axis (Ox) and its potential energy is given by $V(x) = k\dfrac{x^2}{2}$, where k is a positive constant.

1) Show that there is a position that corresponds to a stable equilibrium.
2) We propose to study the movement of M near the equilibrium position, write the equation of movement.

Exercise 2.3

It is pleasant that the suspension of a car gives it a period to which the body is accustomed (the period of walking) is $T_p = 0.8$ (s). We take $g = 10$ (N/m).
1) By schematizing the car by a mass M on a spring of stiffness k.

  - Calculate how much it will lower when we introduce a mass $m = 70$ (kg).
2) Explain why a truck is not comfortable, especially when empty.

Exercise 2.4

In order to compare the own pulsations of free linear oscillators at 1 dof, we study three simple harmonic oscillators (a), (b) and (c).

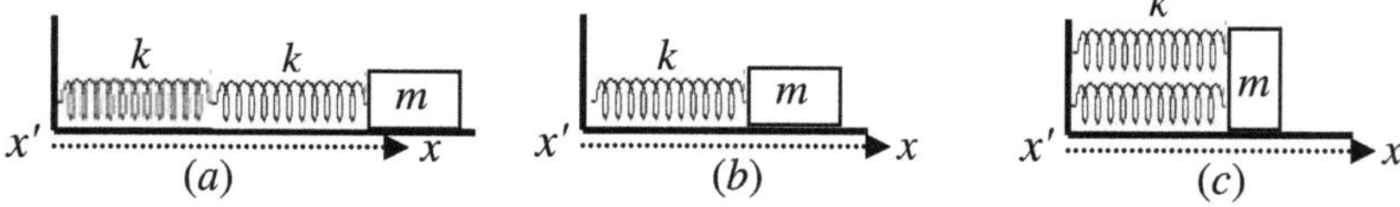

- Show that the ratios of the own pulsations (taken squared ω_0^2) are given by:

$$\frac{\omega_{0(a)}^2}{\omega_{0(b)}^2} = \frac{1}{2}, \quad \frac{\omega_{0(a)}^2}{\omega_{0(c)}^2} = \frac{1}{4} \quad \text{and} \quad \frac{\omega_{0(b)}^2}{\omega_{0(c)}^2} = \frac{1}{2}.$$

Exercise 2.5

A liquid pendulum consists of a vertical U-shaped tube, with a small and uniform section S, containing a liquid of volumic mass ρ. The pressure at the free surface is equal to atmospheric pressure.
At rest, the level is the same in both branches of the tube. When the equilibrium is broken, the liquid oscillates.
- Calculate the own period of the liquid oscillations.

Exercise 2.6

We consider the oscillatory mechanical system rolling without slip of a disk of mass M and radius R, when the disk rotates by θ, its center of gravity moves by x (such that $x = R\theta$).

We take $OA = \dfrac{R}{2}$ and the inertia moment of the disk $J = \dfrac{1}{2}MR^2$.

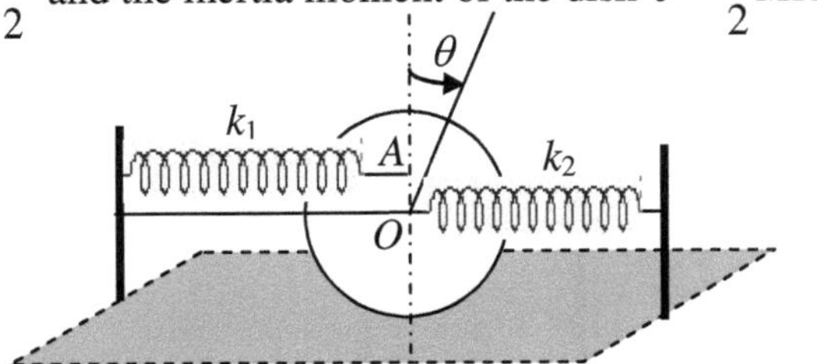

1) Establish the Lagrangian of the system.
2) Determine the differential equation of motion in terms of θ.
3) Deduce the period of the oscillations in the case $k_1 = k_2 = k$.

Exercise 2.7

We consider a harmonic oscillator defined by a mass m placed in an elastic potential $E_p = \dfrac{1}{2}k x^2$ and subjected to a viscous friction force whose coefficient is α.
1) In the case of small free oscillations ($\alpha = 0$), determine the equation for own pulsations and deduce the own frequencies of the oscillations.
2) We admit that friction exists and we see that after an integer p ($p \in N$) the amplitude decreases to a third.

- Give the expression for the displacement, then calculate the logarithmic decrement as well as the friction coefficient α. (Indication, we will assume that T of the damped oscillations $\cong$ to the own period T_0).
- What value will it be necessary to add to α to have critical damping.

Exercise 2.8

Consider a spring, of negligible mass and stiffness $k = 16$ (N/m), suspended vertically in the air and carrying an iron plate of mass $m = 100$ (g). We move the mass m slightly away from its equilibrium position, then we release it.
1) Calculate the own period of the oscillations of the system, note it by T_0.
2) The plate is immersed in a liquid and experiences a braking force proportional to its velocity. The damping coefficient of the system is noted σ. Measurements show that the pseudo-period $T = T_0(1+5\ 10^{-3})$, calculate σ.
- Deduce the quality coefficient Q of the system.

Exercise 2.9

The mechanical system in the figure is composed of a mass m attached to the end of a rigid bar, of length L and negligible mass, the second end can vibrate around point O; a spring of stiffness k attached to the ground is connected to the bar at A such that $OA = L/3$, and a damper of constant α acts on the system at point B, such that $OB = 2L/3$.

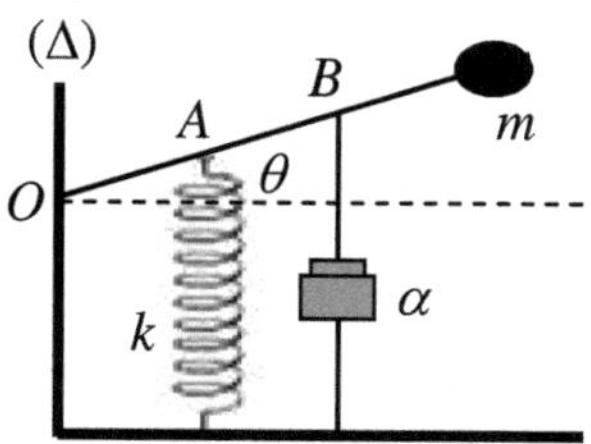

The system can make small movements $\theta(t)$ around an axis (Δ) perpendicular $(\perp)$ in the plane of the figure in O. It is in equilibrium when the bar is horizontal (for $\theta = 0$).
1) Solve the equation of motion for a pseudo-oscillation.
2) Calculate the coefficient α_c for $m = 200$ (g), $k = 20$ (N/m), $L = 18$ (cm).

Exercise 2.10

A material point M of mass m is mobile on a horizontal axis (Ox), and it is subjected to a viscous friction force (of the type $-\alpha\dot{x}$). This point is connected via a spring of stiffness k to a point A of abscissa x_A.

We pose $\omega_0 = \sqrt{\dfrac{k}{m}}$, $\sigma = \dfrac{\alpha}{2m}$ and we assume that $\omega_0 >> \sigma$.

1) What does this hypothesis correspond to?

2) The point A being assumed to be fixed, we move M away from its equilibrium position and we abandon it without initial velocity.

- Calculate the time interval τ at the end of which the amplitude of the movement is divided by e = 2.718...

3) The point is now animated by a movement imposed and described by $x_A = A\cos(\omega t)$. Write the equation of motion of M and look for solutions corresponding to the steady state of the form: $x(t) = C\cos(\omega t + \phi)$ (we use the complex notation).

Exercise 2.11

We consider the following pendulum

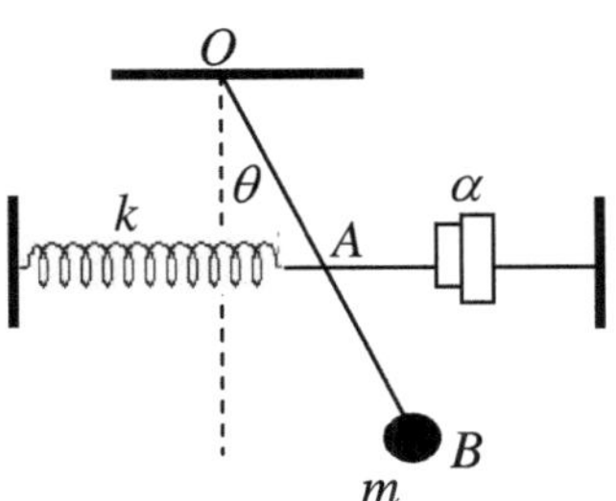

The mass m is punctual. The massless rod OB of length $2L$ pivots around the point O at an angle θ relative to its vertical equilibrium position. The distances $OA = AB = L$.

At rest (for $\theta = 0$), the spring is not deformed. A damping device exerts a fluid friction force at the point A.

1) In the case of weak oscillations, find the differential equation governing the movement of the system.

2) Application: m = 0.1 (kg); k = 4 (N/m); α = 1.2 (kg/s); L = 1(m) and g = 10 (m/s²).

- Calculate the damping coefficient σ, the own pulsation ω_0 and give the corresponding solution (in transient mode).

Exercise 2.12

The motion of an oscillator is described by the following equation:
$$0.2\ddot{x} + 1.2\dot{x} + 5x = 5\cos(2t)$$

1) Determine the own period T_0, the damping coefficient σ and the excitation pulsation ω.

2) Show that the transient solution is a damped oscillatory movement. Deduce its pseudo-pulsation Ω.
- Give the solution, taking into account the initial conditions $x(0) = 0$ and $v(0) = 4$ (m/s).
3) Determine the steady-state solution.

Exercise 2.13

An electron (e) is moving in the electric field of a hydrogenoid ion nucleus, charged $(+Ze)$, Z is the number of protons and e is the absolute value of the (charge of the electron).
1) Find, in polar coordinates, the expression of its kinetic energy $E_c(r, \theta)$ and potential energy $E_p(r, \theta)$, with $x = r \cos(\theta)$ and $y = r \sin(\theta)$.
2) Derive the Lagrangian expression.
3) Write the equations of motion.

Exercise 2.14

Hydrocyanic acid H–C≡N, which is a toxic acid with linear triatomic molecules, is composed of three atoms a-b-c whose nuclei are aligned (a, b and c designate, respectively, the atoms H, C and N). The interaction forces between atoms derive from potentials $V_{ab}(x_{ab})$ and $V_{bc}(x_{bc})$. We note by x_a, x_b et x_c the deviations of atoms from their equilibrium positions and by m_a, m_b and m_c the atomic masses.

1) Determine the fundamental pulsations ω_0 and ω_0' of diatomic molecules ab and bc from interaction potentials $V_{ab}(x_{ab})$ and $V_{bc}(x_{bc})$.
2) Write the coupled equations of motion for the triatomic molecule abc.
3) By changing the variable $x_{ab} = x_b - x_a$ and $x_{cb} = x_b - x_c$, determine the molecule's own vibration pulsations abc.
4) Application with hydrocyanic acid. The own pulsations of H–C≡N measured by spectrometry are $\omega_1 = 6.25 \ 10^{14} \ (s^{-1})$ and $\omega_2 = 3.95 \ 10^{14} \ (s^{-1})$.
- Deduce the fundamental pulsations ω_0 and ω_0' of connections H–C and C≡N.

Exercise 2.15

We consider the assembly composed of two identical cylinders (each of mass M, radius R and inertia moment $J = \dfrac{1}{2}MR^2$), rolling without slipping on a horizontal support.

Let θ_1 and θ_2 be the angles of rotation of these two cylinders relative to their respective equilibrium positions.

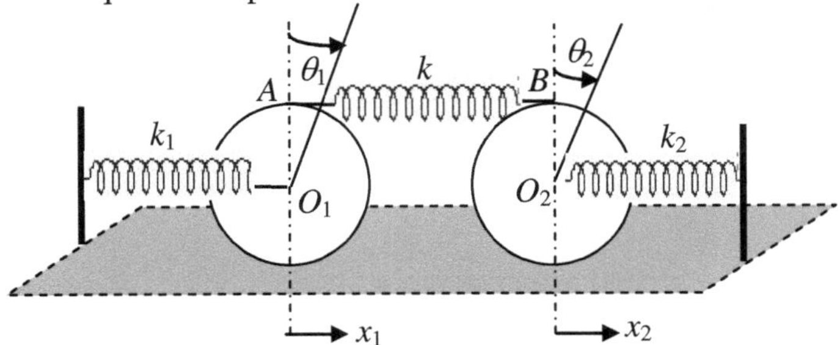

At rest ($\theta_1 = \theta_2 = 0$), the springs are not deformed.
1) Determine the Lagrangian of the system according to x_1 and x_2.
2) We suppose that $k_1 = k_2 = k' \neq k$, find the equations of motion.
3) Deduce the own pulsations.
4) Write the Lagrangian in the following form

$$L = \frac{3}{4} M \left[\dot{x}_1^2 + \dot{x}_2^2 - \omega_0^2 (x_1^2 + x_2^2 - 2kx_1 x_2) \right],$$

- Deduce the expressions of ω_0^2 and K.

5) K is the coupling coefficient, show that it varies between two limit values.

6) Give the meaning of ω_0 comparing it to the pulsations found in question 3).

- Deduce the effect of the K coupling on the own pulsations.

Exercise 2.16

Let us consider a fixed support (S), in the supposed Galilean terrestrial frame of reference.
In the system considered, we locate the positions of the two masses by their abscissa x_1 and x_2 from their equilibrium position.
Data: $k_1 = 78 \; 10^{-3}$ (N/m), $k_2 = 15 \; 10^{-3}$ (N/m), $k_3 = 6 \; 10^{-3}$ (N/m), $m_1 = 3$ (g) and $m_2 = 10^{-3}$ (Kg).
It is assumed that the springs are perfectly elastic and of negligible mass.
It is assumed that the masses cannot move only parallel to the y-axis.
1) Determine the system's own pulsations.
2) Determine $y_1(t)$ and $y_2(t)$.

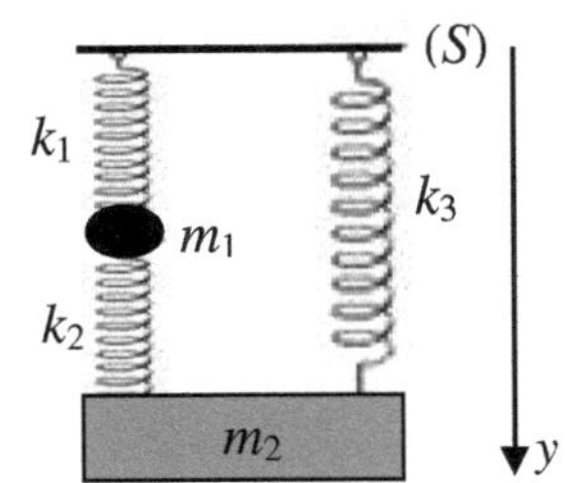

Exercise 2.17

Let consider the mechanical system of the figure.

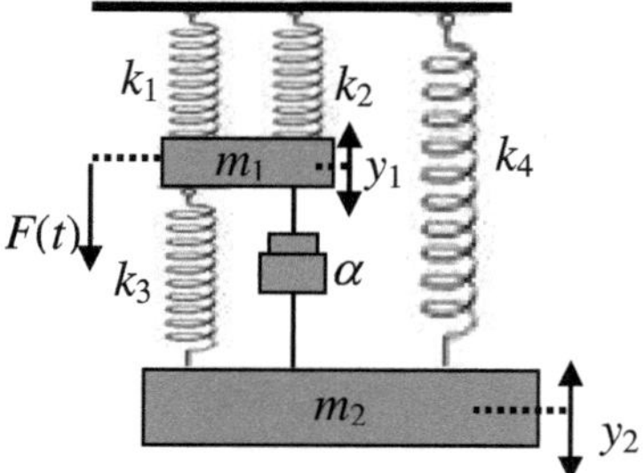

1) Write the equations of free linear oscillations of the masses m_1 and m_2.
2) We assume that the system describes forced damped oscillations such that $F(t) = F_0 \cos(\omega t)$.

2.1) Calculate input impedance $Z_e = \dfrac{F(t)}{\dot{x}_1(t)}$.

Put it in the form $Z_e = Z_1 + \dfrac{Z_2 Z_3}{Z_2 + Z_3}$.

2.2) Propose an electrical diagram equivalent to the mechanical system.

Exercise 2.18

Two identical masses m_1 and m_2 are connected to the frame by two springs of stiffness k and, between them, by a damper of coefficient α, so as to oscillate vertically.

Using two linear combinations $Y_1(t)$ and $Y_2(t)$ of the displacements of the masses m_1 and m_2, determine:

1) The two decoupled differential equations
   in $Y_1(t)$ and $Y_2(t)$ which govern the movement.
2) The system's own pulsations.
3) The general solutions $y_1(t)$ and $y_2(t)$ of the movement
   of the two masses.
4) Completely determine this movement:
   Knowing that at $t = 0$, $y_1(0) = y_0$; $y_2(0) = y_0$
   and $\dot{y}_1(0) = \dot{y}_2(0) = 0$.

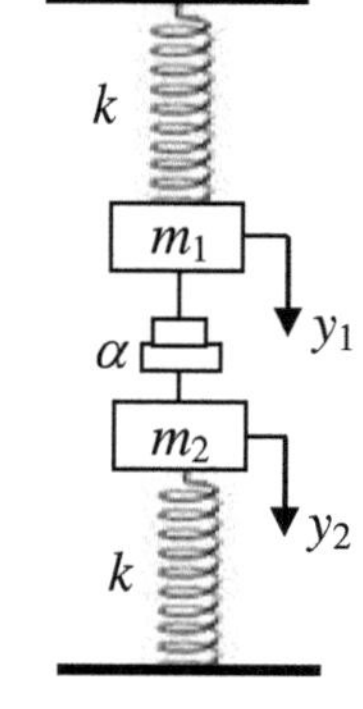

5) Give an example of initial conditions such that the two masses perform free damped oscillations in opposition to phase and pseudo-pulsation Ω.

Exercise 2.19

Consider the oscillating mechanical system, where x_1 and x_2, respectively, designate the positions of the masses m_1 and m_2, relative to their equilibrium positions and the exciting force $F(t)$ is applied only on m_1.

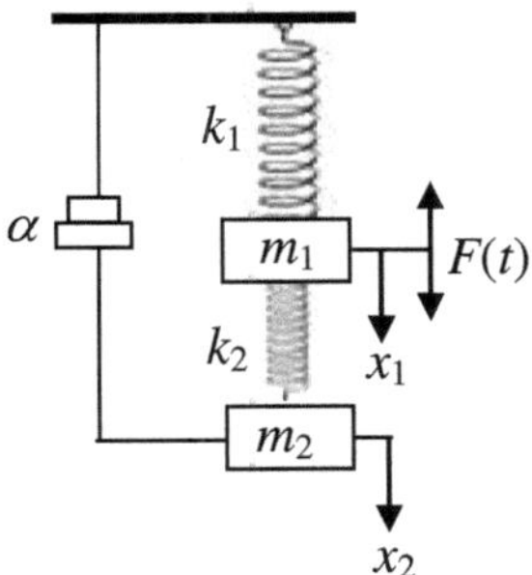

1) Write the equations of motion, in the case $m_1 = m_2 = m$ and $k_1 = k_2 = k$.
2) Seek steady state solutions knowing that $F(t) = k\,a\cos(\omega t)$.

3) If $\alpha = 0$, so that what value of ω do we have resonance?
- Give in this case the condition for which the 1st mass remains immobile.

Exercise 2.20

We consider the double pendulum formed of two masses m_1 and m_2 welded at the end of two lengths L_1 and L_2 of negligible masses.
The position of the system is marked by the angles θ_1 and θ_2.

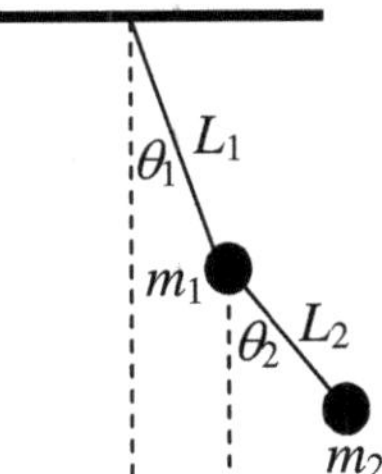

1) Determine the kinetic and potential energies of the system,
2) Deduce the differential equations of motion.

Exercise 2.21

We consider the system in the figure (coupling by spring and shock absorber).

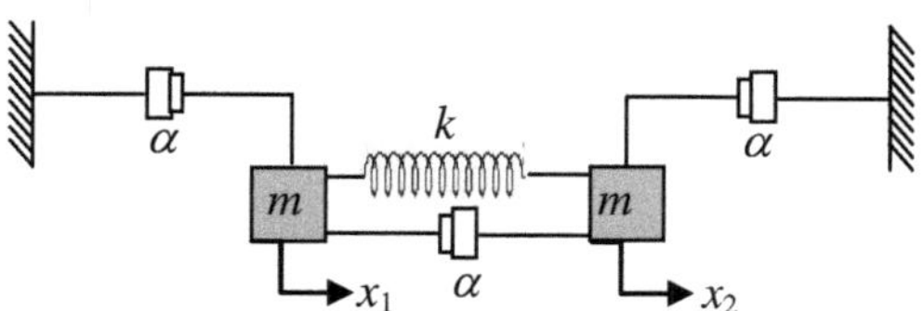

1) Determine the Lagrangian of the free system.
2) From the free Lagrangian, give the equations of the damped system.
3) Solve the system of equations.

Exercise 2.22

We consider a linear atomic chain that we assimilate to a mechanical system composed of N masses and $N+1$ (identical) springs. We note x_1, x_2, x_3, ..., x_N, the displacements of the N masses.
1) Write the potential and kinetic energy of the entire system.
2) Write the Lagrange equations.
3) What happens to this system of differential equations for sinusoidal solutions. We pose $\omega_0 = \sqrt{\dfrac{k}{m}}$.

4) To have the specific pulsations of the system, we end up with a writing in the form $\det[D(u)] = 0$. We ask to explcit u, then, what is the rank of the matrix D.

Answer key to exercise series n° 2

Exercise 2.1

1) The equation of the trajectory (in the form of an ellipse) is $\dfrac{x^2}{a^2}+\dfrac{y^2}{b^2}=1$.

$\begin{cases} x = a\cos(\theta) \\ y = b\sin(\theta) \end{cases}$, two coordinates are required (2 dof system).

Note

If $a = b \equiv$ radius (it's a circle) $\Rightarrow$ in this case you only need one coordinate (which is θ).

2) Case of the cylinder, we need a single coordinate (x or an angle θ).

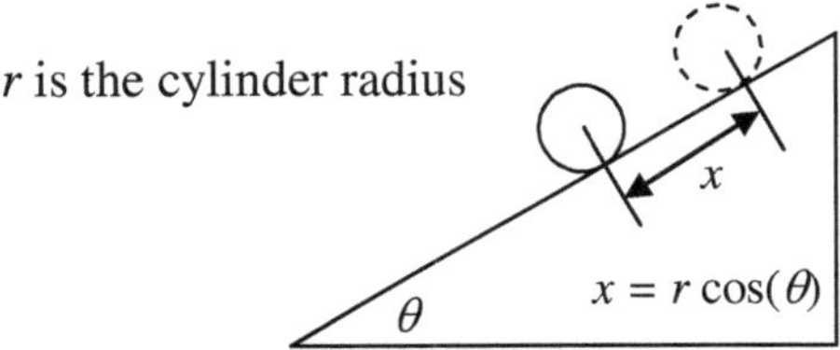

If mass (m) is negligible translational movement only.

If $m \neq 0$, there is translation and rotation.

3) In the case of a simple pendulum, we need only one coordinate (θ) $\Rightarrow$ system one (1) dof.

4) Case of the double pendulum 2 coordinates are required $\Rightarrow$ System 2 dof.

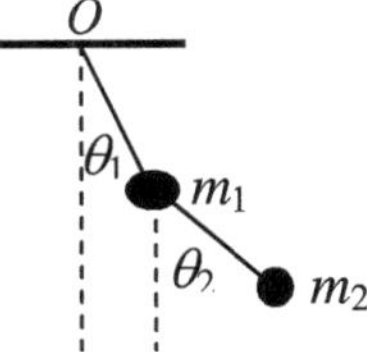

5) *i*) we need two coordinates (x and θ) $\Rightarrow$ system 2 dof.

   ii) you also need two coordinates (θ_1 and θ_2) $\Rightarrow$ system 2 dof.

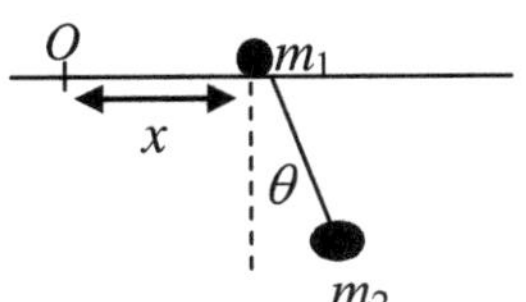

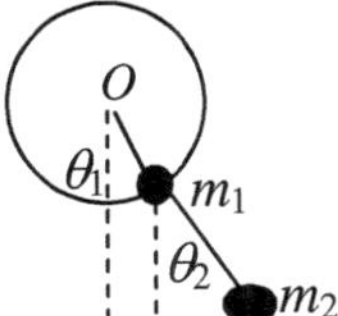

6) We have a system with one (1) dof if the mass M of the pulley is negligible.
If $M \neq 0 \Rightarrow$ the system becomes 2 dof.

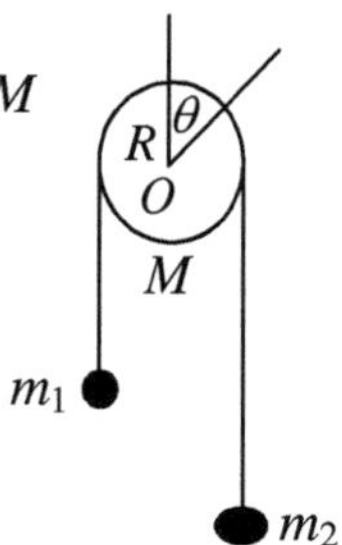

Exercise 2.2

1) To observe the equilibrium positions, we must trace the curve $V(x)$.
Definition domain: the function $V(x)$ is defined on the interval $]-\infty,+\infty[$
Its derivative is $V'(x) = kx$.

It takes zero for $V'(x) = 0 \Rightarrow x = 0$.

Limits $\begin{cases} x \to -\infty \Rightarrow V(x) \to +\infty \\ x \to +\infty \Rightarrow V(x) \to +\infty \end{cases}$

If $x = 0 \Rightarrow V(0) = 0$,

Variation table

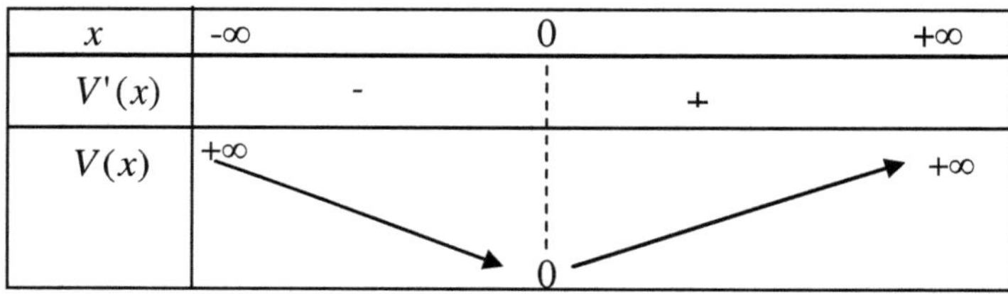

x	$-\infty$		0		$+\infty$
$V'(x)$		$-$		$+$	
$V(x)$	$+\infty$	$\searrow$	0	$\nearrow$	$+\infty$

The curve of the function

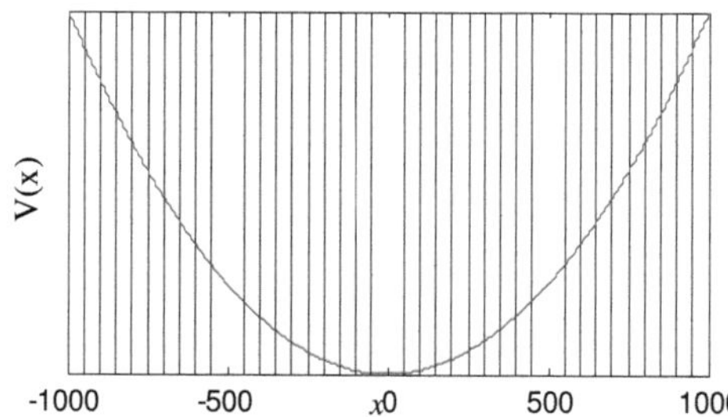

At the neighborhood of $x = 0 \Rightarrow$ the curve presents a minimum (stable equilibrium).

2) The equation of motion

The Lagrangien $L = E_c - E_p = \dfrac{1}{2} m \dot{x}^2 - \dfrac{1}{2} k x^2$.

The equation $\dfrac{d}{dt}\left(\dfrac{\partial L}{\partial \dot{x}}\right) - \dfrac{\partial L}{\partial x} = 0 \Rightarrow \ddot{x} + \dfrac{k}{m} x = 0$.

Exercise 2.3

1) When we add a mass m to the car of mass M, it lowers by a distance x and its period is $T_p = 0.8$ (s).

We know that $T_p = \dfrac{2\pi}{\omega_0}$.

- The own pulsation ω_0 of a system mass-spring $(M\text{-}k)$ is $\omega_0^2 = \dfrac{k}{M}$,

Application $\dfrac{k}{M} = \dfrac{4\pi^2}{T_p^2} \Rightarrow k = \dfrac{4\pi^2 M}{T_p^2} = \dfrac{4\pi^2 1500}{(0.8)^2} = 92433.75$ (N/m)

- On the other hand, the mass m is placed on the plate M connected to the spring. The mass m is in equilibrium; the elastic force compensates its weight

$$P = m g = k x \Rightarrow x = \dfrac{m g}{k} = \dfrac{70 \times 10}{92433.75} = 7.5 \; 10^{-3} \text{ (m)} = 7.5 \text{ (mm)}.$$

The car lowers by 7.5 (mm) when mass m is added.

2) In the case of a truck mass m_c

Its period $T_c = 2\pi \sqrt{\dfrac{m_c}{k_c}} \Rightarrow T_c$ is proportional to m_c.

If m_c increase (with the load), the period T_c increase or slow (which is not practical for comfort).

If the truck is empty, the period T_c is short and the response is quick.

Hence, the unpleasant feeling. Justification by calculations

$k_c = \dfrac{4\pi^2 m_c}{T_p^2}$, if m_c icrease, k_c increase also (equivalent to an iron plate welded to a rigid rod).

The lowering distance if we introduce the mass $m = 70$ (kg), into a truck, is

$x = \dfrac{m g}{k_c} = 0$ (because k_c is very high).

Therefore, the relief of the roadway will be felt (absence of comfort).

Exercise 2.4

It has been demonstrated in several applications that the equation of motion in the case of a mass m subjected to the elastic force of a spring of stiffness constant k (mass-spring system) is $m\ddot{x} + kx = 0$.

- <u>In the case (a)</u>

The two springs can be replaced by a single spring of constant

$\dfrac{1}{k_a} = \dfrac{1}{k} + \dfrac{1}{k} \Rightarrow k_a = \dfrac{k}{2}$. Its equation of motion is as follows

$m\ddot{x} + \dfrac{k}{2}x = 0$, which gives the following own pulsation $\omega_{0(a)}^2 = \dfrac{k}{2m}$.

- <u>In the case (b)</u>

The equation of motion is $m\ddot{x} + kx = 0$, which gives the following own

pulsation $\omega_{0(b)}^2 = \dfrac{k}{m}$.

- <u>In the case (c)</u>

Here the two springs can be replaced by a spring of constant $k_c = k + k = 2k$.

The equation of motion is $m\ddot{x} + 2kx = 0$, which gives the following own

pulsation $\omega_{0(c)}^2 = \dfrac{2k}{m}$.

Let's do the pulsation report

$* \quad \dfrac{\omega_{0(a)}^2}{\omega_{0(b)}^2} = \dfrac{k}{2m} \times \dfrac{m}{k} = \dfrac{1}{2} \; ; \; \dfrac{\omega_{0(a)}^2}{\omega_{0(c)}^2} = \dfrac{k}{2m} \times \dfrac{m}{2k} = \dfrac{1}{4} \; ; \; \dfrac{\omega_{0(b)}^2}{\omega_{0(c)}^2} = \dfrac{k}{m} \times \dfrac{m}{2k} = \dfrac{1}{2}.$

Exercise 2.5

In the case of the liquid pendulum, it is
the weight $\vec{P}$ of the liquid column (grayed)
which produces the restoring force $\vec{F}$.
Therefore, the force F results from force
of gravitation. In the tube, section
right, the levels are moved by x
($-x$ in one branch and $+x$ in the other),
We write $F = -2\rho g S x$.

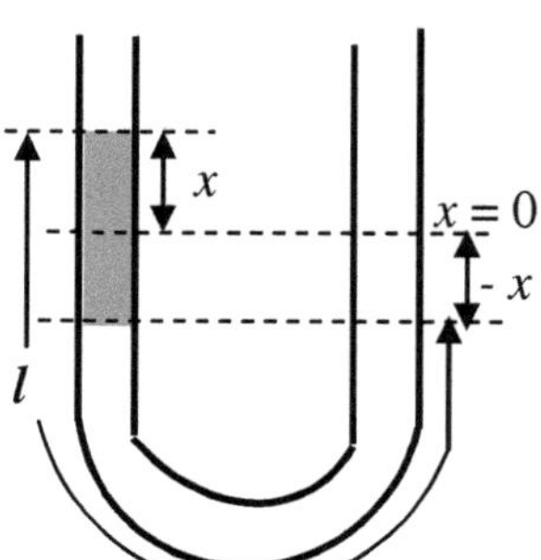

The Lagrangian of the system is given by $L = E_c - E_p$

Since the section is uniform $\Rightarrow$ all the particles making up the fluid have the same velocity (which is the derivative of the displacement x), then the kinetic energy of the fluid system is $E_c = \dfrac{1}{2}m\dot{x}^2$.

The potential energy of the fluid is $E_p = \rho g S x^2$.

The Lagrange's function $L = \dfrac{1}{2}m\dot{x}^2 - \rho g S x^2$.

The equation of motion is given by the Lagrange equation:

$$\frac{d}{dt}\left(\frac{\partial L}{\partial \dot{x}}\right) - \frac{\partial L}{\partial x} = 0.$$

The derivatives give $\dfrac{d}{dt}\left(\dfrac{\partial L}{\partial \dot{x}}\right) = m\ddot{x}$ and $\dfrac{\partial L}{\partial x} = 2\rho g S x$.

The differential equation of the fluid is $\ddot{x} + \dfrac{2\rho g S}{m}x = 0$.

We pose $\omega_0^2 = \dfrac{2\rho g S}{m} \Rightarrow$ the own period is $T_0 = 2\pi\sqrt{\dfrac{m}{2\rho g S}}$.

Exercise 2.6

There are two simultaneous movements of the disc, rotational (angle θ) and translational (distance x), with $x = R\theta$.

When spring k_1 extends by $\dfrac{R}{2}\sin(\theta) + x = \dfrac{R}{2}\sin(\theta) + R\theta$, the spring k_2 is compressed by $x = R\theta$, and vice versa.

The kinetic energy of the disk is $E_c = E_{c-translational} + E_{c-rotational}$.

$$E_c = \frac{1}{2}(\frac{1}{2}MR^2)\dot{\theta}^2 + \frac{1}{2}M\dot{x}^2 = \frac{1}{2}(\frac{1}{2}MR^2)\dot{\theta}^2 + \frac{1}{2}MR^2\dot{\theta}^2$$

The potential energy is $E_p = E_{pk1} + E_{pk2}$.

$$E_p = \frac{1}{2}k_1(x + \frac{R}{2}\sin(\theta))^2 + \frac{1}{2}k_2 x^2 = \frac{1}{2}k_1(R\theta + \frac{R}{2}\sin(\theta))^2 + \frac{1}{2}k_2(R\theta)^2.$$

The Lagrange function is $L = E_c - E_p$.

$$L = \frac{1}{2}(\frac{1}{2}MR^2)\dot{\theta}^2 + \frac{1}{2}MR\dot{\theta}^2 - \frac{1}{2}k_1(R\theta + \frac{R}{2}\sin(\theta))^2 - \frac{1}{2}k_2(R\theta)^2$$

$$\Rightarrow L = \frac{3}{4}MR^2\dot{\theta}^2 - \frac{1}{2}k_1 R^2(\theta + \frac{\sin(\theta)}{2})^2 - \frac{1}{2}k_2 R^2\theta^2.$$

The Lagrange's equation is $\dfrac{d}{dt}\left(\dfrac{\partial L}{\partial \dot\theta}\right)-\dfrac{\partial L}{\partial \theta}=0$.

$$\Rightarrow \frac{6}{4}MR^2\ddot\theta+k_1 R^2(\theta+\frac{\sin(\theta)}{2})(1+\frac{\cos(\theta)}{2})+k_2 R^2\theta=0.$$

In the case of low amplitude oscillations $\sin(\theta)\approx\theta$ and $\cos(\theta)\approx 1$

$$\Rightarrow \ddot\theta+[\frac{9k_1+4k_2}{6M}]\theta=0.$$

We pose $\omega_0^2=\dfrac{9k_1+4k_2}{6M}$, and if $k_1=k_2=k\Rightarrow T=2\pi\sqrt{\dfrac{6M}{13k}}$.

Exercise 2.7

1) The equation of motion, in the case where $\alpha=0$

The Lagrangian $L=E_c-E_p=\dfrac{1}{2}m\dot x^2-\dfrac{1}{2}kx^2$.

Equation of motion $\dfrac{d}{dt}\left(\dfrac{\partial L}{\partial \dot x}\right)-\dfrac{\partial L}{\partial x}=0\Rightarrow m\ddot x+kx=0$. Which can be put in

the classic form $\ddot x+\omega_0^2 x=0$, with $\omega_0^2=\dfrac{k}{m}\Rightarrow T_0=2\pi\sqrt{\dfrac{m}{k}}$.

The oscillation frequency is $f=\dfrac{1}{T_0}=\dfrac{1}{2\pi}\sqrt{\dfrac{k}{m}}$.

2) The equation of motion, in the case where $\alpha\neq 0$

Lagrange's equation $\dfrac{d}{dt}\left(\dfrac{\partial L}{\partial \dot x}\right)-\dfrac{\partial L}{\partial x}=\sum f\Rightarrow m\ddot x+kx=-\alpha\dot x$,

We divide on the mass m, the equation takes the following form

$\ddot x+2\sigma\dot x+\omega_0^2 x=0$, with $2\sigma=\dfrac{\alpha}{m}$.

Expression of displacement is expressed by the solution of the equation
$x(t)=Ae^{-\sigma t}\cos(\omega_0 t+\varphi)$.

    i) The logarithmic decrement

$$DL=\delta=\ln\left(\frac{x(t)}{x(t+pT)}\right)=\ln\left(\frac{Ae^{-\sigma t}\cos(\omega_0 t+\varphi)}{Ae^{-\sigma(t+pT)}\cos(\omega_0(t+pT)+\varphi)}\right)=\ln(e^{\sigma pT})=\sigma pT$$

On the other hand: the amplitudes decrease to a third $\Rightarrow \delta=\ln(3)$.

Finally, $\sigma \, pT = \ln(3)$. Since $\sigma = \dfrac{\alpha}{2m}$, therefore, $\dfrac{\alpha}{2m} \, pT = \ln(3)$

$$\Rightarrow \alpha = \frac{2m\ln(3)}{pT} \approx \frac{2m\ln(3)}{pT_0}.$$

Considering the expression $T_0 = 2\pi\sqrt{\dfrac{m}{k}}$, we find $\alpha = \dfrac{2m\ln(3)}{p\,2\pi\sqrt{\dfrac{m}{k}}} = \dfrac{\ln(3)}{\pi\,p}\sqrt{k\,m}$.

ii) Critical damping

In the case of critical damping $\sigma_c = \omega_0$

We replace $\sigma = \dfrac{\alpha_c}{2m} = \omega_0 = \sqrt{\dfrac{k}{m}} \Rightarrow \alpha_c = 2\sqrt{km}$.

The quantity to add is

$$\Delta\alpha = \alpha_c - \alpha = 2\sqrt{km} - \frac{\ln(3)}{\pi\,p}\sqrt{km},$$

$$= \left[2 - \frac{\ln(3)}{\pi\,p}\right]\sqrt{km}.$$

Exercise 2.8

1) The own period of the oscillations
We first look for the equation of vibrational motion

- For the displacement x, $E_c = \dfrac{1}{2}m\dot{x}^2$,

- Reference $E_p = 0$, is taken at the plane passing through the mass m at equilibrium), and we can assume that the spring is initially extended by a distance x_0

$$E_p = \frac{1}{2}k(x + x_0)^2 - m\,g\,x + cst$$

Equilibrium condition $\left.\dfrac{\partial E_p}{\partial x}\right|_{x=0} = 0$,

The derivative of the potential energy with respect to x gives

$\dfrac{\partial E_p}{\partial x} = k(x + x_0) - m\,g$. For $x = 0 \Rightarrow k\,x_0 - m\,g = 0$ (equilibrium condition).

Taking into account the equilibrium condition, the potential energy simplifies as follows $E_p = \dfrac{1}{2}kx^2 + \underbrace{k\,x\,x_0 - m\,g\,x}_{equil\,condition} + cste = \dfrac{1}{2}kx^2 + cst$.

The Lagrangian of the system $L = \dfrac{1}{2}m\dot{x}^2 - \dfrac{1}{2}kx^2 + cst$.

The equation of motion $\dfrac{d}{dt}\left(\dfrac{\partial L}{\partial \dot{x}}\right) - \dfrac{\partial L}{\partial x} = 0 \Rightarrow m\ddot{x} + kx = 0$.

We divide on $m \Rightarrow \ddot{x} + \dfrac{k}{m}x = 0 \Leftrightarrow \ddot{x} + \omega_0^2 x = 0$, with $\omega_0^2 = \dfrac{k}{m} = (\dfrac{2\pi}{T_0})^2$.

The own period $T_0 = 2\pi\sqrt{\dfrac{m}{k}} = 2\pi\sqrt{\dfrac{0{,}1}{16}} \approx 0.5\,(s)$.

2) The mass m is introduced into a viscous fluid

In this case m is subjected to a braking force $\vec{f} = -\alpha\vec{v} \equiv -\alpha\,\dot{x}\vec{i}$.

The new equation of motion is $\dfrac{d}{dt}\left(\dfrac{\partial L}{\partial \dot{x}}\right) - \dfrac{\partial L}{\partial x} = f$,

$\Rightarrow m\ddot{x} + \alpha\,\dot{x} + kx = 0$. (We write it in the classical form),

$\ddot{x} + 2\sigma\,\dot{x} + \omega_0^2 x = 0$, with $\sigma = \dfrac{\alpha}{2m}$ and $\omega_0^2 = \dfrac{k}{m}$.

We speak of a pseudo-period $\Rightarrow$ the movement is pseudo-periodic.

Its pseudo-pulsation is $\Omega^2 = \omega_0^2 - \sigma^2$.

$\Rightarrow \sigma = \sqrt{\omega_0^2 - \Omega^2} = 2\pi\sqrt{\dfrac{1}{T_0^2} - \dfrac{1}{T^2}}$.

Given that $T = T_0\,(1 + 5\ 10^{-3})$, we replace it in the previous expression:

$$\sigma = \dfrac{2\pi}{T_0}\sqrt{1 - \dfrac{1}{(1+5\ 10^{-3})^2}} = \dfrac{2\pi}{0.5}\sqrt{1 - \dfrac{1}{(1+5\ 10^{-3})^2}} = 0.88\,(s^{-1}).$$

- The quality coefficient

$$Q = \dfrac{\omega_0}{2\sigma} = \dfrac{1}{2}\times\dfrac{2\pi}{T_0}\times\dfrac{T_0}{2\pi\sqrt{1 - \dfrac{1}{(1+5\ 10^{-3})^2}}} = 7.13.$$

Note

The larger Q is, the better the system is.

Exercise 2.9

1) Writing the equation of motion

i) First, we write the equation of motion of the free system ($\alpha = 0$)

We are interested in small movements (θ small)

The displacement of the mass m is $y = L\theta \Rightarrow E_c = \dfrac{1}{2}mL^2\dot\theta^2$,

We take the reference $E_p = 0$, at the horizontal plane containing the point O and the mass m at equilibrium and we can assume that the spring is compressed by a distance y_0

$$E_p = \frac{1}{2}k(y_1 + y_0)^2 + mg\,y = \frac{1}{2}k(\frac{L}{3}\theta + y_0)^2 + mg\,L\theta$$

Equilibrium condition $\left.\dfrac{\partial E_p}{\partial \theta}\right|_{\theta=0} = 0$,

The derivative of the potential energy with respect to θ give

$$\frac{\partial E_p}{\partial \theta} = \frac{L}{3}k(\frac{L}{3}\theta + y_0) + m\,gL.$$

For $\theta = 0 \Rightarrow \dfrac{kL}{3}y_0 + m\,gL = 0$ (equilibrium condition).

Taking into account the equilibrium condition, the potential energy simplifies as follows

$$E_p = \frac{1}{2}k\frac{L^2}{9}\theta^2 + \underbrace{\left[\frac{kLy_0}{3} + mg\,L\right]}_{equil\ \ condition}\theta + cst = \frac{1}{2}k\frac{L^2}{9}\theta^2 + cst.$$

The system's Lagrangian is $L = E_c - E_p = \dfrac{1}{2}mL^2\dot\theta^2 - \dfrac{1}{2}k\dfrac{L^2}{9}\theta^2 + cst$

The equation of motion is obtained from $\dfrac{d}{dt}\left(\dfrac{\partial L}{\partial \dot\theta}\right) - \dfrac{\partial L}{\partial \theta} = 0.$

$$\Rightarrow\ mL^2\ddot\theta + k\frac{L^2}{9}\theta = 0 \ \text{(equation of free oscillations).}$$

ii) In the case of damped oscillations ($\alpha \neq 0$), the corresponding equation is

$$mL^2\ddot\theta + k\frac{L^2}{9}\theta = f = -\alpha\dot y_2.\ \text{Here,}\ y_2 = \frac{2L}{3}\theta.$$

Its final form is $mL^2\ddot{\theta} + \alpha\dfrac{2L}{3}\dot{\theta} + k\dfrac{L^2}{9}\theta = 0$ (equation of damped motion).

It can be written in the classical form $\ddot{\theta} + 2\sigma\dot{\theta} + \omega_0^2\theta = 0$, with $2\sigma = \dfrac{2\alpha}{3mL}$

and $\omega_0^2 = \dfrac{k}{9m}$.

2) The search for the critical coefficient α_c

We first seek the solution of the differential equation of damped motion.

We go through the characteristic equation $r^2 + 2\sigma r + \omega_0^2 = 0$,

$\Delta' = \sigma^2 - \omega_0^2$,

So that he has cushioned solutions $\Rightarrow \Delta' \le 0$.

That's to say $\dfrac{\alpha^2}{9m^2L^2} \le \dfrac{k}{9m} \Rightarrow \alpha \le \sqrt{mkL^2}$.

The critical case corresponds to the condition $\Delta' = 0$: and the coefficient in this case is called critical damping.

$\Rightarrow \alpha_c = \sqrt{mkL^2}$. Application: $\alpha_c = \sqrt{0.02 \times 20 \times (0.18)^2} = 0.11$ (s^{-1}).

Exercise 2.10

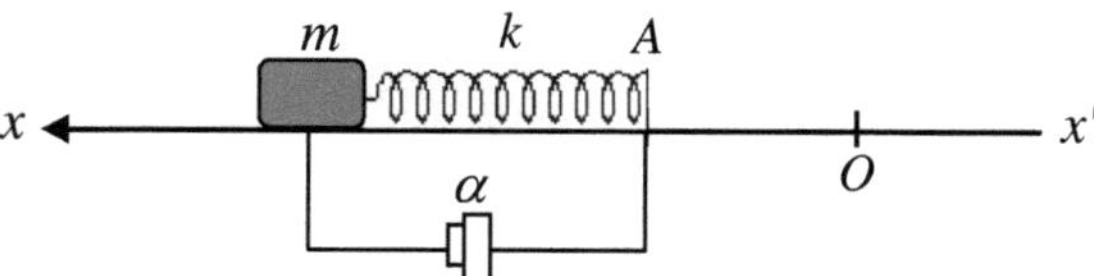

1) Since $\omega_0 \gg \sigma \Rightarrow$ the hypothesis corresponds to the weakly damped regime.

2) The application of the Lagrange formalism makes it easy to find the following differential equation $\ddot{x} + \dfrac{k}{m}x = -\alpha\dot{x}$, which can be written in the

classical form $\ddot{x} + 2\sigma\dot{x} + \omega_0^2 x = 0$, with $\omega_0 = \sqrt{\dfrac{k}{m}}$ and $2\sigma = \dfrac{\alpha}{m}$.

The general solution is of the form $x(t) = Ce^{-\sigma t}\cos(\Omega t + \varphi)$.

Since $\omega_0 \gg \sigma \Rightarrow \Omega \approx \omega_0$, consequently, $x(t) = Ce^{-\sigma t}\cos(\omega_0 t + \varphi)$, C and φ are two constants to be determined using the initial conditions.

The ratio $\dfrac{x(t+\tau)}{x(t)} = \dfrac{C\,e^{-\sigma(\tau+t)}\cos(\Omega(t+\tau)+\varphi)}{C\,e^{-\sigma t}\cos(\Omega t+\varphi)} = e^{-\sigma\tau} = \dfrac{1}{e} \Rightarrow \tau = \dfrac{1}{\sigma}$.

3) The new equation of motion is

$m\ddot{x} + \alpha\,\dot{x} + kx = kx_A = \underbrace{kA}_{F_0}\cos(\omega t)$, knowing that kx_A is a force.

We divide on the mass m, we obtain

$\ddot{x} + 2\sigma\dot{x} + \omega_0^2 x = kx_A = \Gamma\cos(\omega t)$, with $\Gamma = \dfrac{kA}{m}$.

The solution of the equation consists of two parts, solution of the homogeneous equation (without 2nd member) and particular solution (with 2nd member). (To find the two solutions, see the differential equations).

Exercise 2.11

1) The energies (kinetic and potential) of the system

$E_c = \dfrac{1}{2}m(2L\dot{\theta})^2 = 2mL^2\dot{\theta}^2$;

We take as reference $E_p = 0$ (the horizontal plane containing O).

$E_p = \dfrac{1}{2}k(L\theta)^2 - mg2L(1-\cos(\theta))$.

i) We first determine the equation of the free system ($\alpha = 0$)
The Lagrangian of the free system

$L = E_c - E_p = \dfrac{1}{2}m(2L\dot{\theta})^2 = 2mL^2\dot{\theta}^2 - \dfrac{1}{2}k(L\theta)^2 - mg2L\cos(\theta) + cst$.

The equation of motion of the free system is obtained from the following

Lagrange equation $\dfrac{d}{dt}\left(\dfrac{\partial L}{\partial \dot{\theta}}\right) - \dfrac{\partial L}{\partial \theta} = 0$,

$\dfrac{d}{dt}\left(\dfrac{\partial L}{\partial \dot{\theta}}\right) = 4mL^2\ddot{\theta}$ and $\dfrac{\partial L}{\partial \theta} = -kL^2\theta + 2mgL\sin(\theta)$.

Weak amplitude oscillations $\Rightarrow \sin(\theta) \approx \theta$ and $\cos(\theta) \approx 1$.

$4mL^2\ddot{\theta} + (kL^2 - 2mgL)\theta = 0$.

ii) If we introduce the friction force (the equation takes the form)

$4mL^2\ddot{\theta} + L(kL - 2mg)\theta = -\alpha L^2\dot{\theta}$.

2) We can write the equation in classical form $\ddot{\theta} + 2\sigma\dot{\theta} + \omega_0^2\theta = 0$.

With $2\sigma = \dfrac{\alpha}{4m} = \dfrac{1,2}{4\times 0.1} = 3 \Rightarrow \sigma = 1.5\,(\text{s}^{-1})$.

$\omega_0^2 = \dfrac{kL - 2mg}{4mL} = \dfrac{4\times 1 - 2\times 0.1\times 10}{4\times 0.1\times 1} = \dfrac{4-2}{0.4} = 5$ (rd/s).

- Solving the differential equation (with friction)

The characteristic equation is $r^2 + 2\sigma r + \omega_0^2 = 0$,

$\Delta' = \sigma^2 - \omega_0^2 = 1.5^2 - 5^2 = -22.75 \prec 0$.

$\Delta' \prec 0 \Rightarrow$ The movement is damped with a pseudo-pulsation $\Omega = \sqrt{22.75}$.

The general solution is of the form $\theta(t) = Ce^{-\sigma t}\cos(\Omega t + \varphi)$.

According to the data from the exercise $\theta(t) = Ce^{-1.5t}\cos(\sqrt{22.75}\,t + \varphi)$.

C and φ are two constants to be determined using the initial conditions.

Exercise 2.12

We divide the equation on (0.2) and we write the equation in the general form $0.2\ddot{x} + 1.2\dot{x} + 5x = 5\cos(2t) \Rightarrow \ddot{x} + 6\dot{x} + 25x = 25\cos(2t)$.

By identification with the classical form ($\ddot{x} + 2\sigma\dot{x} + \omega_0^2 x = \Gamma\cos(\omega t)$),

We deduce $m = 0.2$ (kg) et $\alpha = 1.2$ and $2\sigma = \dfrac{\alpha}{m} = 6 \Rightarrow \sigma = 3$.

For the period, we have $\omega_0^2 = 25 \Rightarrow \omega = 5 = \dfrac{2\pi}{T_0} \Rightarrow T_0 = 1.25$ (s).

The excitatory pulsation is $\omega = 2$ (rd/s).

2) The transient solution corresponds to the homogeneous solution (without 2nd member)

The characteristic equation is $r^2 + 6r + 25 = 0$,

$\Delta' = \sigma^2 - \omega_0^2 = 9 - 25 = -16 \prec 0 \Rightarrow$ The characteristic equation admits two complex roots (one conjugate of the other),

$r_1 = -3 + j\sqrt{16}$ and $r_1 = -3 - j\sqrt{16}$.

The pseudo-pulsation is $\Omega = 4$ (rd/s).

The homogeneous solution is therefore $x = Ae^{-3t}\cos(4t + \phi)$.

Using the initial conditions, we determine A and ϕ:

$x(0) = A\cos(\phi) = 0 \Rightarrow$ since $A \neq 0$, necessarily $\phi = \dfrac{\pi}{2}$.

The derivative $\dot{x} = -3Ae^{-3t}\cos(4t+\dfrac{\pi}{2}) - 4Ae^{-3t}\sin(4t+\dfrac{\pi}{2})$.

At $t = 0$, $\dot{x}(0) = -4A\sin(\dfrac{\pi}{2}) = 4 \Rightarrow A = -1$.

Ultimately, the homogeneous solution $x = -e^{-3t}\cos(4t+\dfrac{\pi}{2})$.

3) The steady-state solution corresponds to the particular solution.
We choose a solution in the form of the 2nd member (in complex writing)
We pose $x_p = C\cos(2t) = C\exp(j2t+\varphi)$, with $C = cst$.

The constant x_p is a solution $\Rightarrow$ it verifies the global equation of motion
$\ddot{x} + 6\dot{x} + 25x = 25\cos(2t) = 25\exp(j2t)$.

We replace $[-C4 + 12jC + 25C]e^{j\varphi} = 25$.

$$C = \frac{25}{\sqrt{21^2 + 12^2}} = 1 \text{ and } \varphi = \left(\tan(\frac{12}{21})\right)^{-1} = -0.5.$$

The final writing of the particular solution is $x_p = \cos(2t - 0.5)$.

Exercise 2.13

A hydrogenoid is a single-electron ion. It has a structure similar to that of hydrogen. Apart from the electrical charge of the nucleus is (Ze).

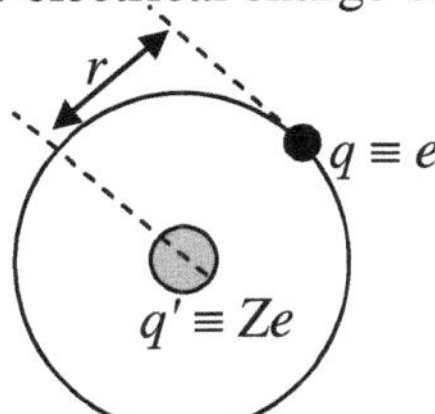

1) The kinetic energy is expressed by $E_c = \dfrac{1}{2}mv^2 = \dfrac{1}{2}m(\dot{x}^2 + \dot{y}^2)$,

In polar coordinates $\begin{cases} x = r\cos(\theta) \\ y = r\sin(\theta) \end{cases} \Rightarrow \begin{cases} \dot{x} = \dot{r}\cos(\theta) - r\dot{\theta}\sin(\theta) \\ \dot{y} = \dot{r}\sin(\theta) + r\dot{\theta}\cos(\theta) \end{cases}$,

By replacing in the expression of E_c, we arrive at $E_c = \dfrac{1}{2}m(\dot{r}^2 + r^2\dot{\theta}^2)$.

The potential energy is given by the integral of the electric force of interaction between the charge of e and the charge of the nucleus (Ze)

$$E_p = -\int F.dr \text{ and } F = \frac{Kqq'}{r^2} = \frac{KZe^2}{r^2} \Rightarrow E_p = -\int \frac{KZe^2}{r^2}.dr = -\frac{KZe^2}{r}.$$

2) The Lagrangian of the system is $L = E_c - E_p = \dfrac{1}{2}m(\dot{r}^2 + r^2\dot{\theta}^2) + \dfrac{KZe^2}{r}$.

3) Lagrange's equations give the equations of motion

$$\begin{cases} \dfrac{d}{dt}\left(\dfrac{\partial L}{\partial \dot{r}}\right) - \dfrac{\partial L}{\partial r} = 0 \rightarrow m\ddot{r} - mr\dot{\theta}^2 + \dfrac{KZe^2}{r^2} = 0, \\[4mm] \dfrac{d}{dt}\left(\dfrac{\partial L}{\partial \dot{\theta}}\right) - \dfrac{\partial L}{\partial \theta} = 0 \rightarrow mr^2\ddot{\theta} + 2mr\dot{\theta} = 0. \end{cases}$$

Exercise 2.14

1) For small deviations from the equilibrium position, the interaction potential between two atoms can be modeled by a spring-like elastic potential.

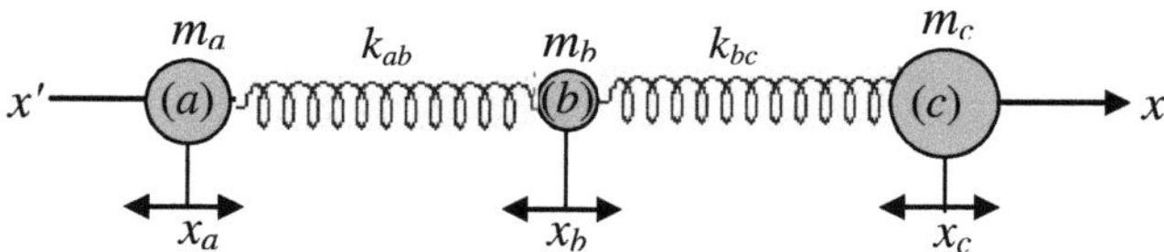

We note by x_a and x_b the displacements of atoms a and b relative to their equilibrium positions. The dynamics equations are written

$$\begin{cases} m_a\ddot{x}_a = k_{ab}(x_b - x_a) \\ m_b\ddot{x}_b = -k_{bc}(x_b - x_c) \end{cases}$$

If we pose $x_{ab} = x_b - x_a$, then, we derivate $\ddot{x}_{ab} = \ddot{x}_b - \ddot{x}_a$,

(We replace the derivatives by their expressions), we end up with

$\ddot{x}_{ab} + \dfrac{k_{ab}}{m_{ab}}x_{ab} = 0$, (here $m_{ab} = \dfrac{1}{m_a} + \dfrac{1}{m_b}$ is the reduced mass of the system).

The own pulsations of the two oscillators (ab) and (bc) are

$$\omega_0 = \sqrt{\dfrac{k_{ab}}{m_{ab}}} = \sqrt{\dfrac{1}{m_{ab}}\left(\dfrac{\partial^2 V_{ab}}{\partial x_{ab}^2}\right)} \quad \text{and} \quad \omega_0' = \sqrt{\dfrac{k_{bc}}{m_{bc}}} = \sqrt{\dfrac{1}{m_{bc}}\left(\dfrac{\partial^2 V_{bc}}{\partial x_{bc}^2}\right)}.$$

2) The equations of motion $\begin{cases} m_a\ddot{x}_a = k_{ab}(x_b - x_a) \\ m_b\ddot{x}_b = -k_{ab}(x_b - x_a) + k_{bc}(x_c - x_b), \\ m_c\ddot{x}_c = -k_{bc}(x_c - x_b) \end{cases}$

The molecule is an isolated system $\Rightarrow \sum_i m_i \ddot{x}_i = 0$, <here $i = a$, b and c (this criterion is verified).

3) By setting $x_{ab} = x_b - x_a$ et $x_{cb} = x_b - x_c$, and dividing on the respective masses, then, making the difference between the equations for C and H on the one hand and N and C on the other hand.

The equations of motion can be put in the form

$$\begin{cases} \ddot{x}_{ab} + \omega_0^2 x_{ab} = -\dfrac{k_{ab}}{m_b} x_{cb} \\[2mm] \ddot{x}_{cb} + \omega_0'^2 x_{cb} = -\dfrac{k_{bc}}{m_b} x_{ab} \end{cases} \Leftrightarrow \begin{cases} \ddot{x}_{HC} + \omega_{0(HC)}^2 x_{HC} + \omega_C^2 x_{CN} = 0 \\[2mm] \ddot{x}_{CN} + \omega_C'^2 x_{CN} + \omega_{0(CN)}^2 x_{HC} = 0 \end{cases}.$$

With $\omega_{0(HC)}^2 = k_{HC}(\dfrac{1}{m_H} + \dfrac{1}{m_C})$, $\omega_{0(CN)}^2 = k_{CN}(\dfrac{1}{m_C} + \dfrac{1}{m_N})$, $\omega_C^2 = \dfrac{k_{CN}}{m_C}$ and $\omega_C'^2 = \dfrac{k_{HC}}{m_C}$.

To have access to the own vibration pulsations of the abc molecule, we must solve the system of equations.

We are looking for harmonic solutions with pulsation ω of type $x_{HC} = X_{HC}e^{j\Omega t}$ and $x_{CN} = X_{CN}e^{j\Omega t}$. In addition, the previous system is written in the following matrix form:

$$\begin{bmatrix} \Omega^2 - \omega_{0(HC)}^2 & \omega_C^2 \\ \omega_C'^2 & \Omega^2 - \omega_{0(CN)}^2 \end{bmatrix} \begin{bmatrix} X_{HC} \\ X_{CN} \end{bmatrix} = \begin{bmatrix} 0 \\ 0 \end{bmatrix}.$$

To have the two own pulsations Ω_1 and Ω_2, we calculate the determinant.

$$\det(D) = \left(\Omega^2 - \omega_{0(HC)}^2\right)\left(\Omega^2 - \omega_{0(CN)}^2\right) - \omega_C^2 \omega_C'^2 = 0$$

$\Omega^4 - (\omega_{0(HC)}^2 + \omega_{0(CN)}^2)\Omega^2 - \omega_C^2\omega_C'^2 = 0$ (Equation of order 4 in Ω, which can be solved by a change of variables).

There are only two physical solutions accepted. They are given by

$$\Omega_1 = \sqrt{\dfrac{1}{2}\left[\omega_{0(HC)}^2 + \omega_{0(CN)}^2 + \sqrt{\left(\omega_{0(HC)}^2 + \omega_{0(CN)}^2\right)^2 + 4\omega_C^2\omega_C'^2}\right]} \quad \text{(antisymmetric mode)}$$

$$\Omega_2 = \sqrt{\dfrac{1}{2}\left[\omega_{0(HC)}^2 + \omega_{0(CN)}^2 - \sqrt{\left(\omega_{0(HC)}^2 + \omega_{0(CN)}^2\right)^2 + 4\omega_C^2\omega_C'^2}\right]} \quad \text{(symmetric mode)}.$$

4) Application

The own pulsations of H–C≡N measured by spectrometry are

$\Omega_1 = \omega_1 = 6.25\ 10^{14}(s^{-1})$ and $\Omega_2 = \omega_2 = 3.95\ 10^{14}\ (s^{-1})$.

- Let's look for the fundamental pulsations $\omega_{0(HC)}$ and $\omega_{0(CN)}'$

Simply take the square of the two expressions of the natural frequencies, (alternatively make the sum/difference between the two equations) then follow the necessary mathematical steps which allow you to have the fundamental pulsations.

Exercise 2.15

1) We have $x_1 = R\theta_1$ and $x_2 = R\theta_2$.

Let us determine the energies (kinetics and potential of the system)

Kinetic energy $E_c = \dfrac{1}{2}M\dot{x}_1^2 + \dfrac{1}{2}J\dot{\theta}_1^2 + \dfrac{1}{2}M\dot{x}_2^2 + \dfrac{1}{2}J\dot{\theta}_2^2 = \dfrac{3}{4}M(\dot{x}_1^2 + \dot{x}_2^2)$,

Potential energy $E_p = \dfrac{1}{2}k_1 x_1^2 + \dfrac{1}{2}k_2 x_2^2 + \dfrac{1}{2}k[(x_1 + R\theta_1) - (x_2 + R\theta_2)]^2$.

We can rewrite it as $E_p = \dfrac{1}{2}k_1 x_1^2 + \dfrac{1}{2}k_2 x_2^2 + 2k(x_1 - x_2)$.

The lagrangian is $L = E_c - E_p$,

$$L = \dfrac{3}{4}M(\dot{x}_1^2 + \dot{x}_2^2) - \dfrac{1}{2}k_1 x_1^2 + \dfrac{1}{2}k_2 x_2^2 + \dfrac{1}{2}k[(x_1 + R\theta_1) - (x_2 + R\theta_2)]^2$$

2) If $k_1 = k_2 = k'$, the Lagrangian becomes

$$L = \dfrac{3}{4}M(\dot{x}_1^2 + \dot{x}_2^2) - \dfrac{1}{2}k'x_1^2 + \dfrac{1}{2}k'x_2^2 + \dfrac{1}{2}k[(x_1 + R\theta_1) - (x_2 + R\theta_2)]^2.$$

The two equations of motion are given by
$$\begin{cases} \dfrac{d}{dt}\left(\dfrac{\partial L}{\partial \dot{x}_1}\right) - \dfrac{\partial L}{\partial x_1} = 0 \\[2mm] \dfrac{d}{dt}\left(\dfrac{\partial L}{\partial \dot{x}_2}\right) - \dfrac{\partial L}{\partial x_2} = 0 \end{cases},$$

After derivation, we obtain
$$\begin{cases} \dfrac{3}{2}M\ddot{x}_1 + (k' + 4k)x_1 - 4kx_2 = 0 \\[2mm] \dfrac{3}{2}M\ddot{x}_2 + (k' + 4k)x_2 - 4kx_1 = 0 \end{cases},$$

If we divide on M, we will have
$$\begin{cases} \ddot{x}_1 + \dfrac{2}{3M}(k' + 4k)x_1 - \dfrac{8k}{3M}x_2 = 0 \\[2mm] \ddot{x}_2 + \dfrac{2}{3M}(k' + 4k)x_2 - \dfrac{8k}{3M}x_1 = 0 \end{cases}.$$

3) We pose solutions of the type $\begin{cases} \tilde{x}_1 = \overline{X}_1 e^{j\omega_0 t} \\ \tilde{x}_2 = \overline{X}_2 e^{j\omega_0 t} \end{cases}$,

$$\Rightarrow \begin{bmatrix} -\omega_0^2 + \dfrac{2}{3M}(k'+4k) & -\dfrac{8k}{3M} \\ -\dfrac{8k}{3M} & -\omega_0^2 + \dfrac{2}{3M}(k'+4k) \end{bmatrix} \begin{bmatrix} \overline{X}_1 \\ \overline{X}_2 \end{bmatrix} = \begin{bmatrix} 0 \\ 0 \end{bmatrix},$$

The calculation of Δ

$$\Delta = [-\omega_0^2 + \frac{2}{3M}(k'+4k)]^2 - [\frac{8k}{3M}]^2$$

$$= [-\omega_0^2 + \frac{2}{3M}(k'+8k)][-\omega_0^2 + \frac{2}{3M}k'] = 0.$$

$\Rightarrow$ The own pulsations are $\begin{cases} \omega_{01}^2 = \dfrac{2}{3M}(k'+8k) \\ \omega_{02}^2 = \dfrac{2}{3M}k' \end{cases}$

4) Expression of the Lagrangian L

$$L = \frac{3}{4}M\left[\dot{x}_1^2 + \dot{x}_2^2 - \omega_0^2(x_1^2 + x_2^2 - 2kx_1 x_2)\right],$$

$$= \frac{3}{4}M\left[\dot{x}_1^2 + \dot{x}_2^2\right] - \frac{1}{2}k'x_1^2 - \frac{1}{2}k'x_2^2 - 2k(x_1 - x_2)^2, \text{ if we develop,}$$

$$L = \frac{3}{4}M\left[\dot{x}_1^2 + \dot{x}_2^2\right] - \frac{1}{2}k'x_1^2 - \frac{1}{2}k'x_2^2 - 2kx_1^2 - 2kx_2^2 + 4kxx_1 x_2,$$

$$L = \frac{3}{4}M\left[\dot{x}_1^2 + \dot{x}_2^2\right] - \frac{1}{2}(k'+4k)x_1^2 - \frac{1}{2}(k'+4k)x_2^2 + 4kx_1 x_2,$$

$$L = \frac{3}{4}M(\dot{x}_1^2 + \dot{x}_2^2 - \frac{2}{3M}(k'+4k)(x_1^2 + x_2^2) + \frac{16}{3M}kx_1 x_2,$$

$$= \frac{3}{4}M[\dot{x}_1^2 + \dot{x}_2^2 - \frac{2}{3M}(k'+4k)(x_1^2 + x_2^2 - \frac{8k}{(k'+4k)}x_1 x_2)],$$

With $\omega_0^2 = \dfrac{2}{3M}(k'+4k)$ and $K = \dfrac{4k}{(k'+4k)}$.

5) $k = 0 \Rightarrow K = 0$, very loose coupling.

$k \to \infty \Rightarrow K = 1$, tight coupling.

So, the coupling coefficient varies between 0 and 1.

6) We see in
$$\begin{cases} \ddot{x}_1 + \dfrac{2}{3M}(k'+4k)x_1 - \dfrac{8}{3M}k\,x_2 = 0 \\[2mm] \ddot{x}_2 + \dfrac{2}{3M}(k'+4k)x_2 - \dfrac{8}{3M}k\,x_1 = 0 \end{cases} \quad \text{and in } \omega_0^2 = \dfrac{2}{3M}(k'+4k) \ .$$

What ω_0 is

> - the pulsation of the first cylinder when the second cylinder is blocked ($x_2 = 0$).
>
> - the pulsation of the second cylinder when the first cylinder is blocked ($x_1 = 0$).

This result shows that the coupling of the two cylinders results in a separation of the two natural pulsations.

$$\omega_{02}^2 = \frac{2k'}{3M} \prec \omega_0^2 = \frac{2}{3M}(k'+4k) \prec \omega_{01}^2 = \frac{2}{3M}(k'+8k).$$

Exercise 2.16

1) It is a 2 dof system, to get the equations, we go through the Lagrangian

- Kinetic energy $E_{c1} = \dfrac{1}{2}m_1\,\dot{y}_1^2 + \dfrac{1}{2}m_2\,\dot{y}_2^2$,

- Potential energy (the ref $E_p = 0$, is taken at the horizontal plane passing the mass m_2 at equilibrium).

$$E_p = \frac{1}{2}k_1\,(y_{01} + y_1)^2 + \frac{1}{2}k_3\,(y_{03} + y_2)^2 + \frac{1}{2}k_2\,(y_{02} + y_2 - y_1)^2$$

$$- m_1\,gy_1 - m_2\,gy_2 + cst$$

(The notations y_{01}, y_{02} and y_{03} are, respectively, the elongations of the springs k_1, k_2 and k_3 at the equilibrium).

Equilibrium conditions $\left(\dfrac{\partial E_p}{\partial y_1}\right)_{y1=0} = 0$ and $\left(\dfrac{\partial E_p}{\partial y_2}\right)_{y2=0} = 0$,

The derivatives are
$$\begin{cases} \dfrac{\partial E_p}{\partial y_1} = k_1\,(y_{01} + y_1) - k_2\,(y_{02} + y_2 - y_1) - m_1\,g \\[3mm] \dfrac{\partial E_p}{\partial y_2} = k_3\,(y_{03} + y_2) + k_2\,(y_{02} + y_2 - y_1) - m_2 g \end{cases}$$

For $y_1 = y_2 = 0 \Rightarrow \begin{cases} 0 = k_1\,y_{01} - k_2\,y_{02} - m_1\,g \\[2mm] 0 = k_3\,y_{03} + k_2\,y_{02} - m_2\,g \end{cases}.$

Taking into account the equilibrium conditions, the rewriting of E_p gives

$$E_p = \frac{1}{2}k_1\,y_1^2 + \frac{1}{2}k_3\,y_2^2 + \frac{1}{2}k_2(y_2 - y_1)^2 + cst\,.$$

The Lagrangian $\dfrac{1}{2}m_1\,\dot{y}_1^2 + \dfrac{1}{2}m_2\,\dot{y}_2^2 - \dfrac{1}{2}k_1\,y_1^2 - \dfrac{1}{2}k_3\,y_2^2 - \dfrac{1}{2}k_2(y_2 - y_1)^2 + cst$

The equations $\begin{cases}\dfrac{d}{dt}\left(\dfrac{\partial L}{\partial \dot{y}_1}\right) - \dfrac{\partial L}{\partial y_1} = 0\\[2mm]\dfrac{d}{dt}\left(\dfrac{\partial L}{\partial \dot{y}_2}\right) - \dfrac{\partial L}{\partial y_2} = 0\end{cases} \Rightarrow \begin{cases}m_1\,\ddot{y}_1 + (k_1 + k_2)y_1 - k_2\,y_2 = 0\\[2mm]m_2\,\ddot{y}_2 + (k_3 + k_2)y_2 - k_2\,y_1 = 0\end{cases}$

We choose solutions in complex notation

$\tilde{y}_1 = \tilde{Y}_1\,e^{j\omega t}$ and $\tilde{y}_2 = \tilde{Y}_2\,e^{j\omega t}$.

The matrix writing of the two equations

$$\begin{bmatrix} -m_1\omega^2 + (k_1 + k_2) & -k_2 \\ -k_2 & -m_1\omega^2 + (k_3 + k_2) \end{bmatrix}\begin{bmatrix}\tilde{Y}_1\\ \tilde{Y}_2\end{bmatrix} = \begin{bmatrix}0\\0\end{bmatrix}.$$

The system admits non-trivial solutions $\Rightarrow$ its determinant is zero.

$\left[-m_1\omega^2 + (k_1 + k_2)\right]\left[-m_1\omega^2 + (k_1 + k_2)\right] - k_2^2 = 0$ (it is an equation witten for the own pulsations),

Application: $3\ 10^{-6}\omega^4 - 156\ 10^{-6}\omega^2 + 1.728\ 10^{-3} = 0$.

The acceptable roots (own pulsations of the system) are $\omega_1 = 4$ (rd/s) and $\omega_2 = 6$ (rd/s).

2) The solutions are given in the form

$$\begin{cases}y_1(t) = A_1\,\cos(4t + \varphi_1) + B_1\,\cos(6t + \theta_2)\\ y_2(t) = A_2\,\cos(4t + \varphi_1) + B_2\,\cos(6t + \theta_2)\end{cases}.$$

Their derivatives are $\begin{cases}\dot{y}_1(t) = -4A_1\,\cos(4t + \varphi_1) - 6B_1\,\cos(6t + \theta_2)\\ \dot{y}_2(t) = -4A_2\,\cos(4t + \varphi_1) - 6B_2\,\cos(6t + \theta_2)\end{cases}$

According to the initial conditions (expressed in cm)

$\begin{cases}1 = A_1\,\cos(\varphi_1) + B_1\,\cos(\theta_2)\\ 0 = A_2\,\cos(\varphi_1) + B_2\,\cos(\theta_2)\end{cases}$ and $\begin{cases}0 = -4A_1\,\cos(\varphi_1) - 6B_1\,\cos(\theta_2)\\ 0 = -4A_2\,\cos(\varphi_1) - 6B_2\,\cos(\theta_2)\end{cases}$

There are several possible solutions.

If, we take $\varphi_1 = \varphi_2 = \theta_1 = 0$ and $\theta_2 = \pi$.

$\Rightarrow A_1 = B_2$ and $A_1 + B_1 = 1$ (then, we report them in the equations)

Therefore $\begin{cases} y_1 = A_1 \cos(4t) + (1 - A_1)\cos(6t) \\ y_2 = A_2\left(\cos(4t) - \cos(6t)\right) \end{cases}$

We find $A_1 = B_2 = 25$ (mm) and $A_2 = 75$ (mm).

Exercise 2.17

1) First, we look for the equivalent system
with $k_{eq} = k_1 + k_2$
We establish the equations of motion
of the free system.
We begin by writing the
expressions of energies.

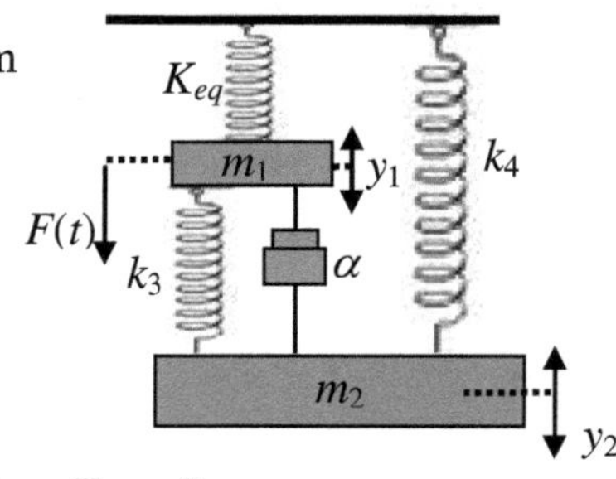

Kinetic energy, $E_c = \dfrac{1}{2}m_1\,\dot{y}_1^2 + \dfrac{1}{2}m_2\,\dot{y}_2^2$,

Potential energies are $E_p = E_{pkeq} + E_{pk3} + E_{pk4} + E_{pm1} + E_{pm2}$.
The reference, $E_p = 0$, is taken from the plane containing the mass m_2.
We assume that the springs undergo an elongation at equilibrium (the k_{eq} is
elongated by y_0, the k_3 by y_{03} and k_4 by y_{04})

$$E_p = \frac{1}{2}k_{eq}(y_0 + y_1)^2 + \frac{1}{2}k_3(y_{03} + y_2 - y_1)^2 + \frac{1}{2}k_4(y_{04} + y_2)^2$$

$$- m_1 g y_1 - m_2 g y_2 + cst.$$

Equilibrium conditions $\left(\dfrac{\partial E_p}{\partial y_1}\right)_{y1=0} = 0$ and $\left(\dfrac{\partial E_p}{\partial y_2}\right)_{y2=0} = 0$.

The derivatives are $\begin{cases} \dfrac{\partial E_p}{\partial y_1} = k_{eq}(y_0 + y_1) - k_3(y_{03} + y_2 - y_1) - m_1 g \\ \dfrac{\partial E_p}{\partial y_2} = k_4(y_{04} + y_2) + k_3(y_{03} + y_2 - y_1) - m_2 g \end{cases}$

For $y_1 = y_2 = 0 \Rightarrow \begin{cases} k_{eq} y_0 - k_3 y_{03} - m_1 g = 0 \\ k_4 y_{04} + k_3 y_{03} - m_2 g = 0 \end{cases}$.

Taking into account the equilibrium conditions, E_p simplifies

$$E_p = \frac{1}{2}k_{eq} y_1^2 + \frac{1}{2}k_4 y_2^2 + \frac{1}{2}k_3(y_2 - y_1)^2 + cst.$$

The Lagrangian $\dfrac{1}{2}m_1\,\dot{y}_1^2 + \dfrac{1}{2}m_2\,\dot{y}_2^2 - \dfrac{1}{2}k_{eq} y_1^2 - \dfrac{1}{2}k_4 y_2^2 - \dfrac{1}{2}k_3(y_2 - y_1)^2 + cst.$

The equations $\begin{cases} \dfrac{d}{dt}\left(\dfrac{\partial L}{\partial \dot{y}_1}\right) - \dfrac{\partial L}{\partial y_1} = 0 \\[3mm] \dfrac{d}{dt}\left(\dfrac{\partial L}{\partial \dot{y}_2}\right) - \dfrac{\partial L}{\partial y_2} = 0 \end{cases} \Rightarrow \begin{cases} m_1\,\ddot{y}_1 + (k_{eq} + k_3)\,y_1 - k_3\,y_2 = 0 \\[2mm] m_2\,\ddot{y}_2 + (k_3 + k_4)\,y_2 - k_3\,y_1 = 0. \end{cases}$

(These are the equations of the free system (without the forces)).

2) When we introduce the forces (friction + force of excitation), the equations of the free system become

$$\begin{cases} m_1\,\ddot{y}_1 + (k_{eq} + k_3)\,y_1 - k_3\,y_2 = F(t) - \alpha(\dot{y}_1 - \dot{y}_2) \\[2mm] m_2\,\ddot{y}_2 + (k_3 + k_4)\,y_2 - k_3\,y_1 = -\alpha(\dot{y}_2 - \dot{y}_1) \end{cases}$$

In steady state, we choose solutions in complex notation

$\tilde{y}_1 = \tilde{Y}_1\, e^{j\omega t}$ and $\tilde{y}_2 = \tilde{Y}_2\, e^{j\omega t}$.

The two equations will take the following form

$$\left(j\omega m_1 + \alpha + \frac{k_{eq} + k_3}{j\omega} \right)\dot{y}_1 + \left(\alpha - \frac{k_3}{j\omega} \right)\dot{y}_2 = F(t) \quad (1)$$

$$\left(j\omega m_2 + \alpha + \frac{k_4 + k_3}{j\omega} \right)\dot{y}_2 + \left(\alpha - \frac{k_3}{j\omega} \right)\dot{y}_1 = 0 \quad\quad (2)$$

From equation (2), we have $\dot{y}_2 = \dfrac{\left(\alpha - \dfrac{k_3}{j\omega} \right)}{\left(j\omega m_2 + \alpha + \dfrac{k_4 + k_3}{j\omega} \right)}\,\dot{y}_1$, we replace in

equation (1)

$$\Rightarrow \left\{ \left(j\omega m_1 + \alpha + \frac{k_{eq} + k_3}{j\omega} \right) + \left(\alpha - \frac{k_3}{j\omega} \right)\frac{\left(\alpha - \dfrac{k_3}{j\omega} \right)}{\left(j\omega m_2 + \alpha + \dfrac{k_4 + k_3}{j\omega} \right)} \right\}\dot{y}_1 = F(t)$$

2.1) By definition, the input impedance is $Z_e = \dfrac{force}{vitesse} \equiv \dfrac{F(t)}{\dot{y}_1}$,

$$Z_e(j\omega) = \left(j\omega m_1 + \alpha + \frac{k_{eq}+k_3}{j\omega} \right) + \left(\alpha - \frac{k_3}{j\omega} \right) \frac{\left(\alpha - \dfrac{k_3}{j\omega} \right)}{\left(j\omega m_2 + \alpha + \dfrac{k_4+k_3}{j\omega} \right)}$$

$$= j\omega m_1 + \frac{k_{eq}}{j\omega} + \left(\alpha + \frac{k_3}{j\omega} \right)\left(1 - \frac{\left(\dfrac{k_{eq}}{j\omega} + \alpha \right)^2}{j\omega m_2 + \alpha + \dfrac{k_4+k_3}{j\omega}} \right),$$

$$= j\omega m_1 + \frac{k_{eq}}{j\omega} + \left(\alpha + \frac{k_3}{j\omega} \right)\left[\frac{j\omega m_2 + \dfrac{k_4}{j\omega}}{j\omega m_2 + \alpha + \dfrac{k_4+k_3}{j\omega}} \right].$$

We pose $Z_1 = j\omega m_1$; $Z_2 = \alpha + \dfrac{k_3}{j\omega}$; $Z_3 = j\omega m_2 + \dfrac{k_4}{j\omega}$,

Finally, $Z_e(j\omega) = Z_1 + \dfrac{Z_2 Z_3}{Z_2 + Z_3}$.

2.2) The equivalent electrical system

We know that $L_1 \Leftrightarrow m_1$ and $L_2 \Leftrightarrow m_2$ (Coils $\Leftrightarrow$ masses),

$R \Leftrightarrow \alpha$ (Resistances $\Leftrightarrow$ friction),

$R \Leftrightarrow \alpha$ (Sources of tension $\Leftrightarrow$ forces of excitation).

$C_{eq} \Leftrightarrow \dfrac{1}{k_1+k_2}$ and $C_3 \Leftrightarrow \dfrac{1}{k_3}$ and $C_4 \Leftrightarrow \dfrac{1}{k_4}$ (Capacitors $\Leftrightarrow$ springs),

The equivalent electrical assembly is as follows

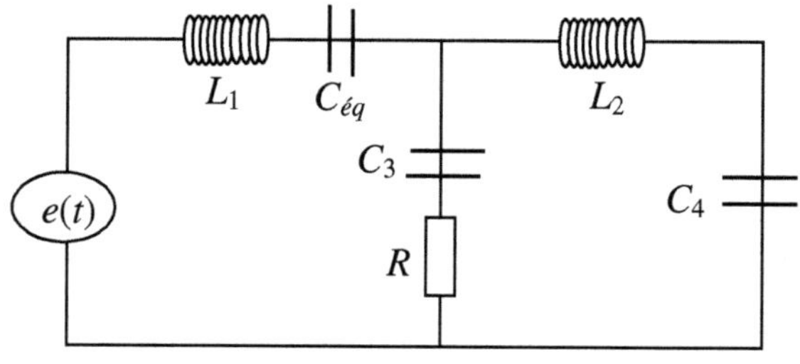

Exercise 2.18

1) With $Y_1 = y_1 + y_2$ and $Y_2 = y_2 - y_1$.

$\Rightarrow$ The decoupled equations are $\begin{cases} m\ddot{Y}_1 + kY_1 = 0 \\ m\ddot{Y}_2 + 2\sigma k\dot{Y}_2 + kY_2 = 0 \end{cases}$, ($Y_1$ and Y_2 are the normal coordinates of the system).

2) The own pulsation $\omega_0 = \sqrt{\dfrac{k}{m}}$, the pseudo-pulsation $\Omega = \sqrt{\omega_0^2 - \sigma^2}$ and $2\sigma = \dfrac{\alpha}{m}$.

3) The solutions are $\begin{cases} y_1(t) = \dfrac{C_1}{2}\sin(\omega_0 t + \varphi_1) + \dfrac{C_2}{2}e^{-\sigma t}\sin(\Omega t + \varphi_2) \\ y_2(t) = \dfrac{C_1}{2}\sin(\omega_0 t + \varphi_1) - \dfrac{C_2}{2}e^{-\sigma t}\sin(\Omega t + \varphi_2) \end{cases}$.

4) With the initial conditions, the two masses oscillate according to mode 1, $\begin{cases} y_1(t) = y_0\cos(\omega t) \\ y_2(t) = y_0\cos(\omega t) \end{cases}$.

5) If at $t = 0$, $\begin{cases} y_1(0) = -y_0 \\ y_2(0) = y_0 \end{cases}$ and $\begin{cases} \dot{y}_1(0) = \sigma y_0, \\ \dot{y}_2(0) = -\sigma y_0. \end{cases}$

$\Rightarrow$ The mode 2 $\begin{cases} y_1(t) = -y_0 e^{-\sigma t}\cos(\Omega t), \\ y_2(t) = y_0 e^{-\sigma t}\cos(\Omega t). \end{cases}$

Exercise 2.19

1) Differential equations: we use the Lagrangian of the system

$$L = \frac{1}{2}m_1\dot{x}_1^2 + \frac{1}{2}m_2\dot{x}_2^2 - \frac{1}{2}k_1 x_1^2 - \frac{1}{2}k_2(x_1 - x_2)^2.$$

The dissipation function is given by $D = \dfrac{1}{2}\alpha\dot{x}_2^2$.

The equations of motion

$$\frac{d}{dt}\left(\frac{\partial L}{\partial \dot{x}_1}\right) - \frac{\partial L}{\partial x_1} = F \quad \text{and} \quad \frac{d}{dt}\left(\frac{\partial L}{\partial \dot{x}_2}\right) - \frac{\partial L}{\partial x_2} = -\frac{\partial D}{\partial \dot{x}_2}.$$

After derivation, we obtain $\begin{cases} m_1\,\ddot{x}_1 + k_1\,x_1 + k_2\,(x_1 - x_2) = F(t) \\ m_2\ddot{x}_2 + k_2\,(x_2 - x_1) = -\alpha\dot{x}_2 \end{cases}$.

Taking into account the fact that $m_1 = m_2 = m$ and $k_1 = k_2 = k$, and dividing by mass, the equations of motion take the following form

$$\begin{cases} \ddot{x}_1 + \dfrac{2k}{m}x_1 - \dfrac{k}{m}x_2 = \dfrac{F(t)}{m} \\[2ex] \ddot{x}_2 + \dfrac{k}{m}x_2 - \dfrac{k}{m}x_1 + \dfrac{\alpha}{m}\dot{x}_2 = 0 \end{cases}$$

2) Steady state solutions

We choose solutions in the form $\begin{cases} \tilde{x}_1 = \bar{x}_1\,e^{j\omega t} \\ \tilde{x}_2 = \bar{x}_2\,e^{j\omega t} \end{cases}$ with $\begin{aligned} \bar{x}_1 &= x_{01}e^{j\varphi_1} \\ \bar{x}_2 &= x_{02}e^{j\varphi_2} \end{aligned}$.

Using the general solutions, we obtain the following matrix writing

$$\underbrace{\begin{bmatrix} -\omega^2 + \dfrac{2k}{m} & -\dfrac{k}{m} \\[2ex] -\dfrac{k}{m} & -\omega^2 + \dfrac{k}{m} + j\omega\dfrac{\alpha}{m} \end{bmatrix}}_{M} \begin{bmatrix} \bar{x}_1 \\[2ex] \bar{x}_2 \end{bmatrix} = \begin{bmatrix} \dfrac{k}{m}a \\[2ex] 0 \end{bmatrix}.$$

To go back to $\bar{x}_1$ and $\bar{x}_2$, we use the determinants of the matrices

$$\det(M) = \left(-\omega^2 + \frac{2k}{m}\right)\left(-\omega^2 + \frac{k}{m} + j\omega\frac{\alpha}{m}\right) - \frac{k^2}{m^2}.$$

To get $\bar{x}_1$, we replace the column of the second member in the first column of the matrix M, then we calculate its determinant again:

$$\det_{(1)} = \det\begin{bmatrix} \dfrac{k}{m}a & -\dfrac{k}{m} \\[2ex] 0 & -\omega^2 + \dfrac{k}{m} + j\omega\dfrac{\alpha}{m} \end{bmatrix} = \left(\frac{ka}{m}\right)\left(-\omega^2 + \frac{k}{m} + j\omega\frac{\alpha}{m}\right),$$

$$\Rightarrow \bar{x}_1 = \frac{\det_{(1)}}{\det(M)} = \frac{\left(\dfrac{ka}{m}\right)\left(-\omega^2 + \dfrac{k}{m} + j\omega\dfrac{\alpha}{m}\right)}{\left(-\omega^2 + \dfrac{2k}{m}\right)\left(-\omega^2 + \dfrac{k}{m} + j\omega\dfrac{\alpha}{m}\right) - \dfrac{k^2}{m^2}}.$$

To get $\bar{x}_2$, we replace the column of the second member in the second column of the matrix M, then we calculate its determinant again

$$\det{}_{(2)} = \det \begin{bmatrix} -\omega^2 + \dfrac{2k}{m} & \dfrac{ka}{m} \\ -\dfrac{k}{m} & 0 \end{bmatrix} = \dfrac{k^2 a}{m^2},$$

$$\Rightarrow \bar{x}_2 = \dfrac{\det{}_{(2)}}{\det(M)} = \dfrac{\dfrac{k^2 a}{m^2}}{\left(-\omega^2 + \dfrac{2k}{m}\right)\left(-\omega^2 + \dfrac{k}{m} + j\omega\dfrac{\alpha}{m}\right) - \dfrac{k^2}{m^2}}.$$

3) Calculation of pulsations in the case where $\alpha = 0$,

$$\det(M)_{\alpha=0} = \left(-\omega^2 + \dfrac{2k}{m}\right)\left(-\omega^2 + \dfrac{k}{m}\right) - \dfrac{k^2}{m^2} \Rightarrow \omega^4 - \dfrac{3k}{m}\omega^2 + \dfrac{k^2}{m^2} = 0.$$

We calculate the roots $\Delta = \dfrac{9k^2}{m^2} - \dfrac{4k^2}{m^2} = \dfrac{5k^2}{m^2} \succ 0$.

The two acceptable solutions are $\omega_{1,2}^2 = \dfrac{1}{2}\left[3\dfrac{k}{m} \pm \dfrac{k}{m}\sqrt{5}\right] = \dfrac{k}{2m}(3 \pm \sqrt{5})$.

- If m_1 is immobile $\bar{x}_1 = 0$ and $\alpha = 0 \Rightarrow \omega^2 + \dfrac{k}{m} = 0$, which gives a pulsation $\omega = \sqrt{\dfrac{k}{m}}$.

Exercise 2.20

1) The kinetic energy of the system is

$$E_c = \dfrac{1}{2}(m_1 + m_2)L_1^2\dot{\theta}_1^2 + \dfrac{1}{2}m_2 L_2^2\dot{\theta}_2^2 + m_2 L_1 L_2\dot{\theta}_1\,\dot{\theta}_2$$

The potential energy of the system is (the reference $E_p = 0$ is taken in the horizontal plane passing through the point of attachment).

$$E_p = -m_1 gL_1\cos(\theta_1) - m_2 g[L_1\cos(\theta_1) + L_2\cos(\theta_2)].$$

2) The Lagrangian of the system is $L = E_c - E_p$

$$L = \dfrac{1}{2}(m_1 + m_2)L_1^2\dot{\theta}_1^2 + \dfrac{1}{2}m_2 L_2^2\dot{\theta}_2^2 + m_2 L_1 L_2\dot{\theta}_1\,\dot{\theta}_2$$

$$+ m_1 gL_1\cos(\theta_1) + m_2 g[L_1\cos(\theta_1) + L_2\cos(\theta_2)]$$

The equations of motion can be obtained easily, by calculating the derivatives of the following equations

$$\frac{d}{dt}\left(\frac{\partial L}{\partial \dot{\theta}_1}\right) - \frac{\partial L}{\partial \theta_1} = 0 \text{ and } \frac{d}{dt}\left(\frac{\partial L}{\partial \dot{\theta}_2}\right) - \frac{\partial L}{\partial \theta_2} = 0.$$

Exercise 2.21

1) The Lagrangian of the system is determined from $L = E_c - E_p$,

- Kinetic energy $E_c = \frac{1}{2} m_1 \dot{x}_1^2 + \frac{1}{2} m_2 \dot{x}_2^2 = \frac{1}{2} m(\dot{x}_1^2 + \dot{x}_2^2)$.

- The potential energy (the reference $E_p = 0$ is taken at the level of the horizontal plan that contains the two masses), $E_p = \frac{1}{2} k(x_2 - x_1)^2$.

$$\Rightarrow L = \frac{1}{2} m(\dot{x}_1^2 + \dot{x}_2^2) - \frac{1}{2} k(x_2 - x_1)^2.$$

2) The equations of motion of the damped system
First, we determine the two equations of the free system

$$\begin{cases} \dfrac{d}{dt}\left(\dfrac{\partial L}{\partial \dot{x}_1}\right) - \dfrac{\partial L}{\partial x_1} = 0 \\[2mm] \dfrac{d}{dt}\left(\dfrac{\partial L}{\partial \dot{x}_2}\right) - \dfrac{\partial L}{\partial x_2} = 0 \end{cases} \Rightarrow \begin{cases} m\ddot{x}_1 + k(x_1 - x_2) = 0 \\[2mm] m\ddot{x}_2 + k(x_2 - x_1) = 0 \end{cases}.$$

If we take into account the shock absorbers, the equations will become:

$$\begin{cases} m\ddot{x}_1 + k(x_1 - x_2) = -\alpha \dot{x}_1 - \alpha(\dot{x}_1 - \dot{x}_2) \\[2mm] m\ddot{x}_2 + k(x_2 - x_1) = -\alpha \dot{x}_2 - \alpha(\dot{x}_2 - \dot{x}_1) \end{cases}$$

3) To solve the system of equations, we choose solutions in the following form (we use the complex notation),

We pose $\begin{cases} x_1(t) = \tilde{X}_1 \exp(j\omega t) \\[2mm] x_2(t) = \tilde{X}_2 \exp(j\omega t) \end{cases}$

Then, we replace in the two equations of motion of the damped system, we obtain the following matrix writing

$$\underbrace{\begin{bmatrix} -m\omega^2 + k + 2\alpha\omega j & -(k + \alpha\omega j) \\[2mm] -(k + \alpha\omega j) & -m\omega^2 + k + 2\alpha\omega j \end{bmatrix}}_{D} \begin{bmatrix} \tilde{X}_1 \\[2mm] \tilde{X}_2 \end{bmatrix} = \begin{bmatrix} 0 \\[2mm] 0 \end{bmatrix}.$$

For the system to have non-trivial solutions, the determinant of the matrix D must be zero.

$$\det(D) = \underbrace{\left(-m\omega^2 + k + 2\alpha\omega j\right)^2}_{A^2} - \underbrace{\left(-k - \alpha\omega j\right)^2}_{B^2} = 0$$

We recall that $A^2 - B^2 = (A + B) \times (A - B)$.

By introducing the new writing and dividing by the mass m, we arrive at:

$$\det(D) = \left[-\omega^2 + \frac{2k}{m} + \frac{3\alpha}{m}\omega j\right] \times \left[-\omega^2 + \frac{\alpha}{m}\omega j\right] = 0.$$

Either
$$\begin{cases} -\omega^2 + \dfrac{2k}{m} + \dfrac{3\alpha}{m}\omega j = 0 & (I) \\[3mm] -\omega^2 + \dfrac{\alpha}{m}\omega j = 0 & (II) \end{cases}$$

De $(II) \Rightarrow \omega(-\omega + \frac{\alpha}{m} j) = 0$, we have two solutions $\omega = 0$ and $\omega = \frac{\alpha}{m} j$.

De $(I) \Rightarrow -\omega^2 + \frac{2k}{m} + \frac{3\alpha}{m}\omega j = 0$, (equation of 2nd degree).

Its discriminant $\Delta = \left(\frac{3\alpha}{m} j\right)^2 + 4\left(\frac{2k}{m}\right) = \frac{-9\alpha^2}{m^2} + \frac{8k}{m} = j^2\left(\frac{9\alpha^2}{m^2} - \frac{8k}{m}\right).$

Discussion

1) if $\Delta > 0 \Rightarrow \omega_3 = \dfrac{\dfrac{3\alpha}{m} j + j\sqrt{\dfrac{-9\alpha^2}{m^2} + \dfrac{8k}{m}}}{2}$ and $\omega_4 = \dfrac{\dfrac{3\alpha}{m} j - j\sqrt{\dfrac{-9\alpha^2}{m^2} + \dfrac{8k}{m}}}{2}.$

2) if $\Delta = 0 \Rightarrow \omega_3 = \omega_4 = \dfrac{3\alpha}{2m} j.$

3) if $\Delta < 0 \Rightarrow \omega_3 = \dfrac{\dfrac{3\alpha}{m} j + j\sqrt{\dfrac{9\alpha^2}{m^2} - \dfrac{8k}{m}}}{2}$ and $\omega_3 = \dfrac{\dfrac{3\alpha}{m} j - j\sqrt{\dfrac{9\alpha^2}{m^2} - \dfrac{8k}{m}}}{2}.$

Note
- For the 1st vibration mode $\omega = 0$, there is no vibration (no movement).

Demonstration

If $\omega = 0$, the corresponding matrix is $\begin{bmatrix} \dfrac{k}{m} & -\dfrac{k}{m} \\ -\dfrac{k}{m} & \dfrac{k}{m} \end{bmatrix} \begin{bmatrix} \tilde{X}_1 \\ \tilde{X}_2 \end{bmatrix} = \begin{bmatrix} 0 \\ 0 \end{bmatrix}$,

$$\Rightarrow \begin{cases} \dfrac{k}{m}\tilde{X}_1 - \dfrac{k}{m}\tilde{X}_1 = 0 \\ -\dfrac{k}{m}\tilde{X}_1 + \dfrac{k}{m}\tilde{X}_2 = 0 \end{cases}, \text{ which leads to } \tilde{X}_1 = \tilde{X}_2.$$

The solutions are $\begin{cases} x_1(t) = \tilde{X}_1 \exp(j\omega 0) = \tilde{X}_1 \\ x_2(t) = \tilde{X}_2 \exp(j\omega 0) = \tilde{X}_1 \end{cases} \Rightarrow x_1(t) = x_2(t) = cst$.

This means that we have no oscillations.

Exercise 2.22

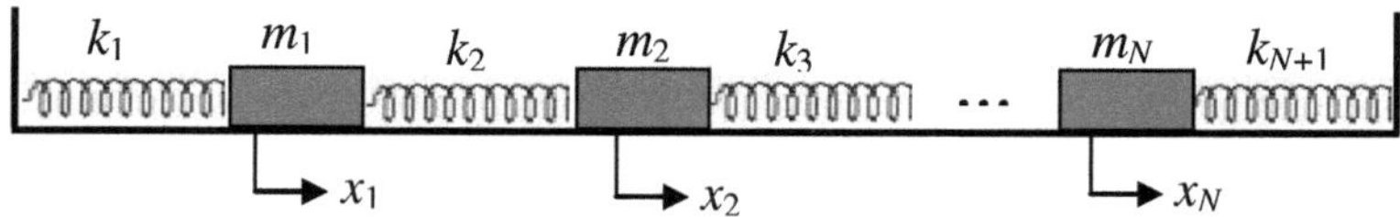

1) The energies (potential and kinetic) of the system

- Kinetic energy, $E_c = \dfrac{1}{2}m_1\dot{x}_1^2 + \dfrac{1}{2}m_2\dot{x}_2^2 + \dfrac{1}{2}m_3\dot{x}_3^2 + ... + \dfrac{1}{2}m_N\dot{x}_N^2$,

- Potential energy (ref(E_p) = 0, the horizontal plane containing the masses:

$$E_p = \frac{1}{2}k_1 x_1^2 + \frac{1}{2}k_2(x_2 - x_1)^2 + \frac{1}{2}k_3(x_3 - x_2)^2 + ... + \frac{1}{2}k_N(x_{N+1} - x_N)^2 + \frac{1}{2}k_{N+1}x_{N+1}^2$$

2) The Lagrange equations are obtained from: $\dfrac{d}{dt}\left(\dfrac{\partial L}{\partial \dot{x}_i}\right) - \dfrac{\partial L}{\partial x_i} = 0$.

In total, there are N equations

Equation (1) $\rightarrow m_1\ddot{x}_1 + k_1 x_1 + k_2(x_2 - x_1) = 0$, (for the mass m_1).

Equation (2) $\rightarrow m_2\ddot{x}_2 + k_2(x_2 - x_1) + k_3(x_2 - x_3) = 0$, (for the mass m_2).

Equation (3) $\rightarrow m_3\ddot{x}_3 + k_3(x_3 - x_2) + k_4(x_3 - x_4) = 0$, (for the mass m_3).

...

Equation (n) $\rightarrow m_n\ddot{x}_n + k_n(x_n - x_{n-1}) + k_{n+1}(x_n - x_{n+1}) = 0$, (for the mass m_n).

...

Equation (N) $\rightarrow m_N\ddot{x}_N + k_N(x_N - x_{N-1}) + k_{N+1}x_N = 0$, (for the mass m_N).

We can rewrite the N equations in the following form

$$m_1 \ddot{x}_1 + (k_1 + k_2)x_1 - k_2 x_2 = 0,$$

$$m_2 \ddot{x}_2 + (k_2 + k_3)x_2 - k_2 x_1 - k_3 x_3 = 0,$$

$$m_3 \ddot{x}_3 + (k_3 + k_4)x_3 - k_3 x_2 - k_4 x_4 = 0,$$

$$\cdots$$

$$m_n \ddot{x}_n + (k_n + k_{n+1})x_n - k_n x_{n-1} - k_{n+1} x_{n+1} = 0,$$

$$\cdots$$

$$m_N \ddot{x}_N + (k_N + k_{n+1})x_N - k_N x_{N-1} = 0.$$

Particular case

We consider the case where the masses and springs are identical
$(m_1 = m_2 = \ldots = m_N = m \quad \text{et } k_1 = k_2 = \ldots = k_N = K_{N+1} = k)$.

3) The system solutions (in the particular case)

We pose $x_n(t) = A_n \cos(\omega t + \varphi_n)$ and $\omega_0 = \dfrac{k}{m}$.

After replacement in the N differential equations, we obtain

$$(\omega^2 - 2\omega_0^2)A_1 + \omega_0^2 A_2 = 0,$$

$$(\omega^2 - 2\omega_0^2)A_2 + \omega_0^2 A_3 + \omega_0^2 A_1 = 0,$$

$$(\omega^2 - 2\omega_0^2)A_3 + \omega_0^2 A_2 + \omega_0^2 A_4 = 0,$$

$$\cdots$$

$$(\omega^2 - 2\omega_0^2)A_n + \omega_0^2 A_{n-1} + \omega_0^2 A_{n+1} = 0,$$

$$\cdots$$

$$(\omega^2 - 2\omega_0^2)A_N + \omega_0^2 A_{N-1} = 0.$$

The N equations can be grouped in the form of a matrix

$$\begin{bmatrix} \omega^2 - 2\omega_0^2 & \omega_0^2 & 0 & \cdots & 0 \\ \omega_0^2 & \omega^2 - 2\omega_0^2 & \omega_0^2 & \cdots & 0 \\ 0 & \omega_0^2 & \omega^2 - 2\omega_0^2 & \cdots & 0 \\ \vdots & \vdots & \vdots & \vdots & \vdots \\ 0 & 0 & 0 & \cdots & \omega^2 - 2\omega_0^2 \end{bmatrix} \begin{bmatrix} A_1 \\ A_2 \\ A_3 \\ \vdots \\ A_N \end{bmatrix} = \begin{bmatrix} 0 \\ 0 \\ 0 \\ \vdots \\ 0 \end{bmatrix}.$$

If, we divide on $\omega_0^2 \Rightarrow$ the matrix takes the new form

$$\begin{bmatrix} u & 1 & 0 & ... & 0 \\ 1 & u & 1 & ... & 0 \\ 0 & 1 & u & ... & 0 \\ \vdots & \vdots & \vdots & \vdots & \vdots \\ 0 & 0 & 0 & ... & u \end{bmatrix} \equiv D, \text{ with } u = \frac{\omega^2 - 2\omega_0^2}{\omega_0^2}.$$

The rank of matrix D is ($N{\times}N$). It allows access to the N roots, which represent the own frequencies of the system.

Chapter 5

Introduction to 1D propagation phenomena

1. Generalities and definitions

- If we have disturbances (oscillations) in time around the equilibrium position (confined in space) $\rightarrow$ we speak of vibration. Purpose of the Part I.
- If the disturbances in time propagate in space $\rightarrow$ we speak of another phenomenon called "waves". Object of the Part II.

Examples

1) Electromagnetic wave is the vibratory movement of electrons in a transmitting or receiving antenna that is at the origin of wave phenomena.
2) Sound wave (also called Acoustic vibration) is a pressure wave created either by the vibration of a membrane (speaker is a vibration of the air) or by the vibration of a string (violin) which propagates.
3) Elastic waves are due to atomic displacements in solids.
4) Spin waves are due to the oscillations of the amplitudes of the spin vectors (spin precession).

Mechanical wave is the phenomenon of propagation of a disturbance in a medium without material transport. The wave carries only energy. We distinguish:
- Transversal waves (shear): during the passage of the disturbance, the material is shortly displaced in a direction perpendicular to the direction of propagation.
- Longitudinal waves (compression): matter is shortly displaced in the direction of propagation.

Note 1

- Remember that a wave is called *a plane wave* if it propagates in a well-defined direction and *monochromatic* if it has a single well-defined frequency.
- A wave propagates, starting from the source, in all the directions that are offered to him.
- The disturbance is transmitted step-by-step (transfer of energy without transport of matter).

- The velocity of propagation of a wave is a property of the medium.
- Two waves can cross each other without disturbing each other.
- The propagation medium of a wave can be 3D (sound wave, light wave), 2D (wave on the surface of water), or 1D (wave on a vibrating string).

2. Propagation equation

For example, a 1D wave propagating from its source is *progressive*. After reflection on an obstacle, this wave propagates in the opposite direction (the wave is *regressive*).

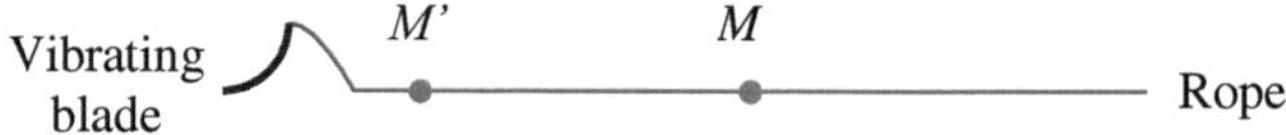

The disturbance at point M at time t is that which previously existed at a point M' at $t' = t - \tau$, with $\tau = M'M/v$, τ being the delay and v the velocity (for non-dispersive media).

La propriété essentielle de cette onde est sa double périodicité (spatiale et temporelle).
The essential property of this wave is its double periodicity (spatial and temporal).
If we call « ψ » the function associated with a displacement of a point of the propagating wave, at a distance "x" from the source, we have:
$\psi(x,t) = \psi(x,t+T) = \psi(x+\lambda,t)$, where T and λ are, respectively, the temporal the spatial periods.

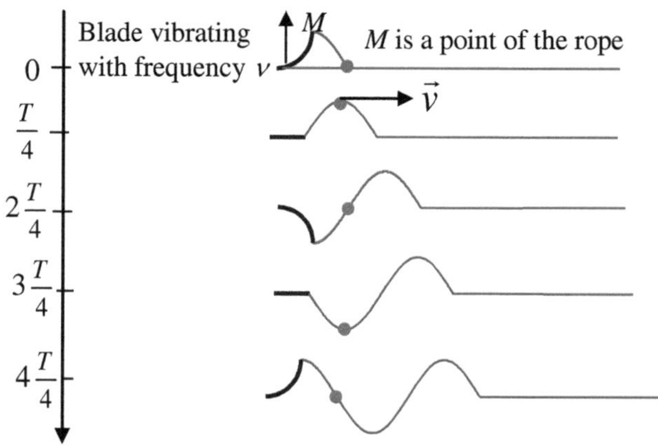

The disturbance (elongation or impulse) propagates at $v = cst$ in the rope. The covered distance $x = v.\Delta t$.

Reminders

1) When the disturbance propagates (going from x to $x' > x$), the function ψ, which describes it can be written as $\psi = f(\underbrace{x - vt}_{x'})$ (waves going to $x > 0$ -

progressives or incidents).

2) If the disturbance propagates towards the negative x ($x' < x$), it is always described by the function ψ, such that $\psi = g(\underbrace{x + vt}_{x'})$ (regressive waves).

3) The disturbance can also be described by the sum of the two functions f and g, $f \pm g$. It means $\psi(x,t) = f(x - vt) \pm g(x + vt)$.

The function ψ is also described as $\psi(x,t) = h(t - \dfrac{x}{v}) \pm g(t + \dfrac{x}{v})$, (dividing

by v).

2.1. Propagation equation

    a) Case of a traveling wave ($\psi(x,t) = f(x - vt)$)

The different points of motion at constant velocity v

$$\frac{\partial \psi}{\partial x} = \frac{\partial \psi}{\partial t} . \frac{\partial t}{\partial x} = \frac{1}{v} . \frac{\partial \psi}{\partial t}$$ (Calculation of the 1st derivative with respect to x).

$$\frac{\partial \psi}{\partial t} = \frac{\partial \psi}{\partial x} . \frac{\partial x}{\partial t} = v . \frac{\partial \psi}{\partial x}$$ (Calculation of the 1st derivative with respect to t).

- Calculation of the 2nd derivative

$$\frac{\partial^2 \psi}{\partial x^2} = \frac{\partial}{\partial x}\left(\frac{\partial \psi}{\partial x}\right) = \frac{1}{v} . \frac{\partial^2 \psi}{\partial x . \partial t} \rightarrow \frac{\partial^2 \psi}{\partial x . \partial t} = v . \frac{\partial^2 \psi}{\partial x^2} ,$$

$$\frac{\partial^2 \psi}{\partial t^2} = \frac{\partial}{\partial t}\left(\frac{\partial \psi}{\partial t}\right) = v . \frac{\partial^2 \psi}{\partial t . \partial x} \rightarrow \frac{\partial^2 \psi}{\partial t . \partial x} = \frac{1}{v} . \frac{\partial^2 \psi}{\partial t^2} . \text{ But, } \frac{\partial^2 \psi}{\partial t . \partial x} = \frac{\partial^2 \psi}{\partial x . \partial t}$$

$$\Rightarrow \frac{\partial^2 \psi(x,t)}{\partial x^2} - \frac{1}{v^2} . \frac{\partial^2 \psi(x,t)}{\partial t^2} = 0 \text{ (Wave propagation equation).}$$

The main characteristic of the progressive wave being that the disturbance is found identical to itself a little further a little later.

Therefore, the propagation medium must be of infinite dimension or of larger dimension compared to the wavelength.

b) Case of a wave propagating in the decreasing x
$$(\psi(x,t) = g(x+vt))$$

Same work as the previous case, and we will find the same equation.

c) Generalization (case of 3D waves)

We write $\psi(x,y,z,t) \equiv \psi(r,t)$, with $r = (x,y,z)$.

The wave propagation equation is given by

$$\frac{\partial^2 \psi(r,t)}{\partial x^2} - \frac{\partial^2 \psi(r,t)}{\partial y^2} - \frac{\partial^2 \psi(r,t)}{\partial z^2} - \frac{1}{v^2}\cdot\frac{\partial^2 \psi(r,t)}{\partial t^2} = 0,$$

which is also written as $\Delta \psi(r,t) - \dfrac{1}{v^2}\cdot\dfrac{\partial^2 \psi(r,t)}{\partial t^2} = 0$,

with $\Delta = (\dfrac{\partial^2}{\partial x^2}, \dfrac{\partial^2}{\partial y^2}, \dfrac{\partial^2}{\partial z^2})$ is the Laplacien.

- We can use the Albertian operator ($\square\,\psi$), defined as

$$\square\,\psi = \Delta \psi(r,t) - \frac{1}{v^2}\cdot\frac{\partial^2 \psi(r,t)}{\partial t^2} = 0.$$

Note 2

The function ψ describes a state of a scalar physical quantity (*elongation of a spring, deviation of a pendulum, etc.*). Nevertheless, a wave state can also be described by a vector quantity (*electric field, magnetic field, etc.*).

2.2. Waves: planes, harmonics and spherical
a) Notion of plane wave (PW) and equiphase surface

The notion of plane wave is widely used in wave theory although the plane wave (PW) does not exist in nature, nor in the laboratory, but its writing presents appreciable conveniences in calculations.

An interesting case for the profile of $\psi(r,t)$ when the functions f and g are sinusoidal (harmonics).

It is written $\psi(r,t) = \psi_0 \cos(\omega t - \vec{k}.\vec{r})$ $\equiv$ of the same shape as

$$\psi_0 \cos[\omega(t - \frac{x}{v})] = \psi_0 \cos(\omega t - \frac{x\omega}{v}).$$

$\vec{k}\begin{pmatrix} k_x \\ k_y \\ k_z \end{pmatrix}$ is the wave vector, and the wavelength is given by $\lambda = \dfrac{2\pi}{k}$.

$\vec{r}$ is the vector position.

$$\psi(\vec{r},t) = \underbrace{\psi_0}_{amplitude} \cos(\underbrace{\omega}_{pulsation} t - \underbrace{k_x.x - k_y.y - k_z.z}_{scalar\ product}) .$$

The term $(\omega t - \vec{k}.\vec{r}) = \phi$ is called the phase of the wave.

At 1D, $k_y = k_z = 0$ and $k_x \neq 0$.

At 2D, $k_z = 0$ and $(k_x, k_y) \neq (0,0)$.

Wave front (equiphase surface)
It is the locus where all the points located on the same abscissa will have a unique phase ϕ.

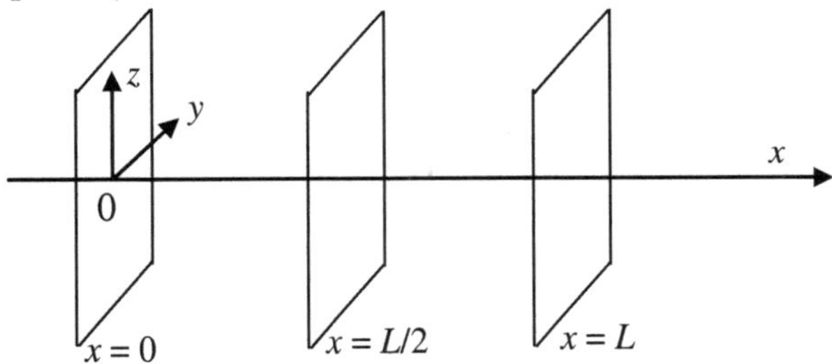

Equiphase plans (of the wave).
These plans vibrate with a delay compared to the source x = 0.
(t = distance L/velocity v).

b) Definition of plane-wave
Wave fronts are planes perpendicular to the direction of the wave.

$\psi(r,t) = \psi_0 \cos(\omega t - \vec{k}.\vec{r})$, the phase $\phi = \omega t - \vec{k}.\vec{r}$, at the time t (fixed)

$\rightarrow k_x.x - k_y.y - k_z.z = cste$, (define the equation of a plane).

- If $k_y = k_z = 0 \Rightarrow k_x.x = cste$, ($x = cst$, equation of a plan // (yOz)).

- If $k_x = k_y = 0 \Rightarrow z = cste$, (equation of a plan // (xOy)).

c) Spherical wave
The wave fronts are spherical surfaces concentric with the source of the wave placed in $r = 0$. Its general equation is

$\psi(r,t) = \psi_0 \cos(\omega t - k.r)$, the phase $\phi = \omega t - k.r$, at the time t (fixed).

$\rightarrow r = Cst$ (equation of a sphere). These waves are emitted by point sources.

2.3. Example of a train composed of two monochromatic waves

Superposition of two plane waves

We consider $\psi_1(\vec{r},t) = \psi_0 \cos(\omega t - \vec{k}.\vec{r})$, $\psi_2(\vec{r},t) = \psi_0 \cos(\omega t + \vec{k}.\vec{r})$,

$$\psi_{tot} = \psi_1 + \psi_2 = \psi_0 \cos(\omega t - \vec{k}.\vec{r}) + \psi_0 \cos(\omega t + \vec{k}.\vec{r})$$

$$\psi_{tot} = 2\psi_0 \cos\left[\left(\frac{\omega_1 + \omega_2}{2}\right)t - \left(\frac{\vec{k}_1 + \vec{k}_2}{2}\right)\times\vec{r}\right]\cos\left[\left(\frac{\omega_1 - \omega_2}{2}\right)t - \left(\frac{\vec{k}_1 - \vec{k}_2}{2}\right)\times\vec{r}\right].$$

The superposition of waves creates interference. The latter can be constructive if the amplitudes of the waves add up, or on the contrary, destructive if the amplitudes are subtracted to the point of giving zero resulting amplitude.

Particular case: stationary (standing) waves

The previous interference takes a particular form in the case where the two superimposed waves are vectors of waves of the same intensities and of opposite directions $\vec{k}_2 = -\vec{k}_1$.

In this case, the sum gives;

Either, $\psi_1(\vec{r},t) = \psi_0 \cos(\omega t - \vec{k}.\vec{r})$,

Or, $\psi_2(\vec{r},t) = \psi_0 \cos(\omega t + \vec{k}.\vec{r})$,

So, $\psi_{tot} = \psi_1 + \psi_2 = 2.\psi_0 \cos(\omega t).\cos(\vec{k}.\vec{r})$.

Conclusion

Adding two progressive monochromatic waves of opposite directions and of the same frequency ω produces a vibration that is not progressive: it is a standing wave.

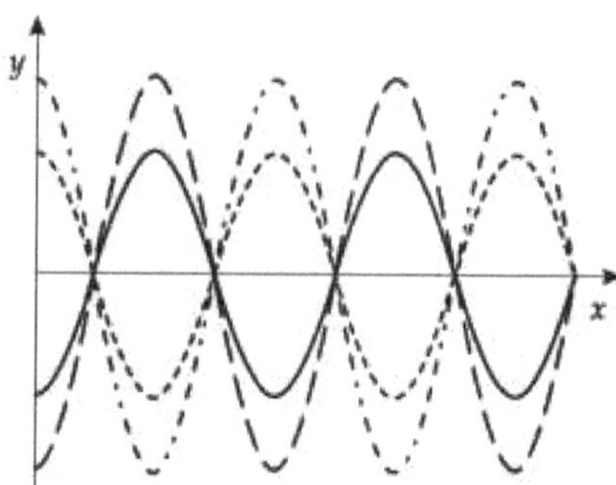

Example of stationary wave at one dimension.

It is seen that the maxima and minima of the amplitude of the standing wave occur at fixed positions.
- The vibration minima of a standing wave designate *the nodes*.
- The vibration maxima of a standing wave designate *the antinodes*.

2.4. Group and phase velocities

The wave that results from the combination of several monochromatic waves is called a wave packet. This wave packet shows an envelope (shape that encompasses all signal fluctuations).

Consider two monochromatic waves of the same amplitude
$$\psi_1 = \psi_0 \cos \omega_0 (\omega_1 t - \vec{k}_1 . \vec{r}) \text{ and } \psi_2 = \psi_0 \cos(\omega_2 t + \vec{k}_2 \vec{r}),$$
The resultant of the sum is given by
$$\psi_{tot} = \psi_1 + \psi_2 = \psi_0 \cos \omega_0 (\omega_1 t - \vec{k}_1 . \vec{r}) + \psi_0 \cos(\omega_2 t + \vec{k}_2 \vec{r}),$$
If the vector $\vec{r}(x, y, z)$ is arbitrary, then:
$$\psi_{tot} = 2.\psi_0 \underbrace{\cos\left[\omega_m t - \vec{k}_m \vec{r}\right]}_{Structure(average\ wave)} \underbrace{\cos\left[\Delta\omega t - \Delta\vec{k}.\vec{r}\right]}_{Envelope}$$

with: $\vec{k}_m = \dfrac{\vec{k}_1 + \vec{k}_2}{2}$, $\Delta\vec{k} = \dfrac{\vec{k}_1 - \vec{k}_2}{2}$, $\omega_m = \dfrac{\omega_1 + \omega_2}{2}$ and $\Delta\omega = \dfrac{\omega_1 - \omega_2}{2}$.

The resulting wave is a product of two cosines. The first cosine defines an average traveling wave of wave vector $\vec{k}_m$ and of pulsation ω_m, the 2nd the 2nd is interpreted as the propagation of an amplitude envelope $2.\psi_0$. In the writing of the sum, we note two velocities:

i) The velocity of the average wave (known as the phase velocity)

It is given by $v_m \equiv v_\varphi = \dfrac{\omega_m}{\left\|\vec{k}_m\right\|}$.

ii) Envelope propagation velocity

The velocity of the envelope is expressed by $v_{envelope} = \dfrac{\Delta\omega}{\left\|\Delta\vec{k}\right\|}$.

If the two wave vectors and the two pulsations are very close to each other, the previous velocity expression tends to *the derivative*.

We then speak of *group velocity*. We write, $v_{envelope} \equiv v_g = \dfrac{d\omega}{\left\| d\vec{k} \right\|}$.

The velocity v_g by definition is the speed with which the information propagates in the medium.

Relation between the two velocities

We have $v_{\varphi} = \dfrac{\omega}{k} \Rightarrow \omega = v_{\varphi}.k$,

and $v_g = \dfrac{d\omega}{dk} = \dfrac{d}{dt}(v_{\varphi}.k) = \dfrac{dk}{dk}.v_{\varphi} + k.\dfrac{dv_{\varphi}}{dk} = v_{\varphi} + k.\dfrac{dv_{\varphi}}{dk}$,

On the other hand, $k = \dfrac{2\pi}{\lambda} \rightarrow \ln(k) = \ln(2\pi) - \ln(\lambda) \Rightarrow \dfrac{dk}{k} = -\dfrac{d\lambda}{\lambda}$

So, $v_g = v_{\varphi} - \lambda \dfrac{dv_{\varphi}}{d\lambda}$, dispersion equation (or Rayleigh criterion).

3. Wave dispersion in a medium
3.1. Definition
We speak of dispersion of a medium if the phase velocity of the wave depends on the frequency. The medium is then said to be dispersive..
3.2. Different type of dispersion
i) Normal dispersion
Normal dispersion is characterized by the increase of the phase velocity v_{φ} when the wavelength λ increases. In other words, the refractive index decreases when the frequency of the wave decreases.

$\dfrac{dv_{\varphi}}{d\lambda} \succ 0$ and $v_g \prec v_{\varphi}$.

In this case the group velocity has the expression $v_g = v_{\varphi} - \lambda \dfrac{dv_{\varphi}}{d\lambda}$.

ii) Abnormal dispersion
The abnormal dispersion is characterized by the decrease in the phase velocity v_{φ} when the wavelength λ increases. In other words, the refractive index increases when the frequency of the wave decreases.

$\dfrac{dv_{\varphi}}{d\lambda} \prec 0$ and $v_g \prec v_{\varphi}$.

iii) No scatter
The phase velocity does not depend on the wavelength.

$$\frac{dv_\varphi}{d\lambda} = 0 \text{ and } v_g = v_\varphi.$$

3.3. Wave propagation in a dispersive medium

If the propagation velocity v of the wave depends on the frequency ω (its wavelength), the medium is said to be dispersive. This means that the refractive index (n) of the medium depends on the frequency of the wave passing through it. We write $v = f(\omega)$ (for a dispersive medium).

Moreover, if $v = cst$, $\forall\ \omega \rightarrow$ we have a non-dispersive medium.

We then introduce two velocities (phase and group velocities), because the propagation takes place in a frequency domain and the representation of this domain is called spectrum.

Note 3

1) To pass from any disturbance to the frequency spectra, the Fourier transform is used.

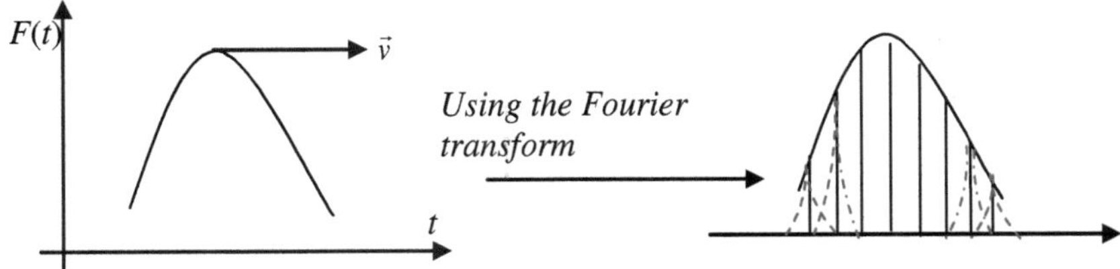

2) A disturbance (impulse or light) is always made up of a sum (discrete or continuous) of elementary waves, because the notion of monochromatic wave (with a single frequency) does not exist either in nature or in the laboratory.

Chapter 6

Wave propagation in a vibrating string; transverse waves

Introduction

Transverse strains only exist in the solid state. There is only one direction of propagation for the wave. There are two types of deformations:
i) Elastic, in this case, the deformation disappears with the removal of the cause (forces, pressure, etc.).
ii) Plastic, in this case, the effect of the deformation remains even after the elimination of the cause that it generated.

1. Vibrating strings and propagation equation

Assumption **1**, it is assumed that the strings studied are one-dimensional (1D) media. In addition, they are non-elastic and characterized by the linear density μ (mass per unit length, its unit is (kg/m)).

Assumption **2**, we consider that the rope is stretched by a tension T and at equilibrium the points of the rope are identified by the abscissa x.
Out of equilibrium, each point $M(x)$ undergoes a transverse disturbance $U(x,t)$ with respect to the direction of the chord.
Consider a strand of the string, subjected to a transverse disturbance, of length $\vec{dl}$ at the time t, and the points A and B are subjected to the tensions $\vec{T}_A$ and $\vec{T}_B$.

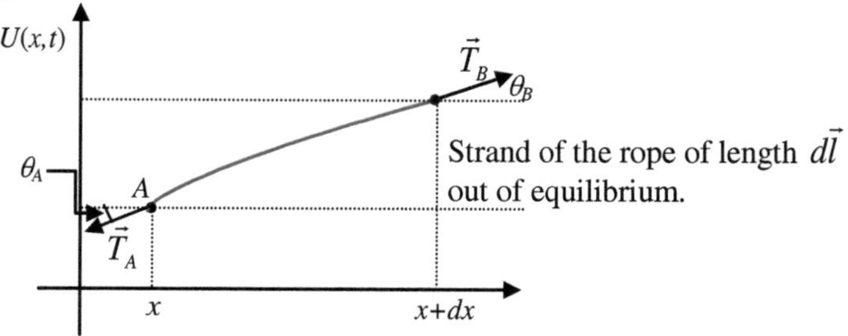

The projection of the tension force vectors gives

$$\vec{T}_A \begin{cases} -T_A \cos(\theta) \\ -T_A \sin(\theta) \end{cases} \text{ and } \vec{T}_B \begin{cases} T_B \cos(\theta) \\ T_B \sin(\theta) \end{cases}.$$

It is assumed that the transverse strains are small $\Rightarrow \sin(\theta) \approx \tan(\theta) \approx \theta$.

By applying the fundamental relation of the dynamics on the strand of the rope, we obtain $\begin{cases} -T\cos(\theta_A) + T\cos(\theta_B) = 0 & (1) \\ -T_A\sin(\theta_A) + T\sin(\theta_B) = \mu\,dx\,a & (2) \end{cases}$, knowing that the movement takes place perpendicularly (in the direction (Oy)) its acceleration $a = \dfrac{\partial^2 U(x,t)}{\partial t^2}$.

From the equation (1) $\Rightarrow \cos(\theta_A) = \cos(\theta_B)$.

We introduce the following notations $\begin{cases} \theta_A = \theta_x \\ \theta_B = \theta_{x+dx} \end{cases}$,

Substituting into equation (2), we get $T\left[\tan(\theta_{x+dx}) - \tan(\theta_x)\right] = \mu\,dx\,\dfrac{\partial^2 U(x,t)}{\partial t^2}$

$$T\left[\left(\dfrac{\partial U}{\partial x}\right)_{x+dx} - \left(\dfrac{\partial U}{\partial x}\right)_x\right] = \mu\,dx\,\dfrac{\partial^2 U(x,t)}{\partial t^2}.$$

On the other hand, $\tan(\theta_x) \equiv slope \equiv \left.\dfrac{\partial U}{\partial x}\right|_x$. And $f(x+dx) = f(x) + \dfrac{\partial f}{\partial x}dx$.

We write then $\left.\dfrac{\partial U}{\partial x}\right|_{x+dx} = \dfrac{\partial U}{\partial x} + \dfrac{\partial^2 U}{\partial x^2}.dx$.

$\Rightarrow T \times dx \times \dfrac{\partial^2 U(x,t)}{\partial x^2} = \mu \times dx \times \dfrac{\partial^2 U(x,t)}{\partial t^2}$, which can be put in the form:

$\dfrac{\partial^2 U(x,t)}{\partial x^2} - \dfrac{\mu}{T}\dfrac{\partial^2 U(x,t)}{\partial t^2} = 0$ (transverse wave propagation equation along the string).

Let's find the dimensions of the ratio $\dfrac{\mu}{T}$

$$\left[\dfrac{\mu}{T}\right] = \dfrac{[\mu]}{[T]} \equiv \dfrac{(m/l)}{(ml/t^2)} \equiv \dfrac{MT^2}{LML} = \left(\dfrac{T}{L}\right)^2 = \dfrac{1}{[velocity]^2} \equiv \dfrac{1}{v^2}.$$

2. Harmonic disturbances

The solution of the wave equation is $U(x,t) = \underset{incident\,wave}{f\left(t - \dfrac{x}{v}\right)} + \underset{reflected\,wave}{g\left(t + \dfrac{x}{v}\right)}$.

It can also be written as $U(x,t) = A\sin[\omega(t - \frac{x}{v})] + B\sin[\omega(t + \frac{x}{v})]$.

Particular case

An interesting case to examine, it corresponds to the situation where the two ends of the rope (of length L) are fixed (rope rigidly taut).

$$\begin{cases} x = 0 \Rightarrow U(0,t) = 0 \\ x = L \Rightarrow U(L,t) = 0 \end{cases}, \text{ then } U(0,t) = A\sin[\omega t] + B\sin[\omega t] \Rightarrow A = -B.$$

And $U(L,t) = A\sin[\omega(t - \frac{L}{v})] + B\sin[\omega(t + \frac{L}{v})] = 0$.

For an intermediate value of x $(0 < x < L)$, the equation $U(x,t)$ is written

$$U(x,t) = A\left\{ \sin[\omega(t - \frac{x}{v})] - \sin[\omega(t + \frac{x}{v})] \right\}.$$

It is equivalent to the expression $U(x,t) = 2A\sin(\frac{\omega x}{v})\cos(\omega t)$ (3).

Nature of spread and propagation

Nature of wave propagation along the rope

From equation (3) and at the end, for $x = L$,

$$U(L,t) = 2A\sin(\frac{\omega L}{v})\cos(\omega t) = 0.$$

As $A \neq 0$ (otherwise no movement) $\Rightarrow$ necessarily: $\sin(\frac{\omega L}{v}) = 0$

$$\Rightarrow \frac{\omega L}{v} = n\pi, \, n \text{ is a natural number.}$$

The final solution is $\omega_n = \frac{n\pi v}{L}$ (called the own pulsation of the string).

Important

The string can vibrate with several modes appropriate to natural frequencies

$$\omega_n \, (f_n = \frac{\omega_n}{2\pi} = \frac{nv}{2L}, \, n \in \aleph).$$

If $n = 1 \rightarrow \omega_1$ is said to be first (1st) harmonic.
If $n = 2 \rightarrow \omega_2$ is said to be 2nd harmonic.
If $n = 3 \rightarrow \omega_3$ is said to be 3rd harmonic.
...

Up to $n \rightarrow \omega_n$ is said to be n^{th} harmonic.

We will therefore have a length $L = \dfrac{n\pi v}{\omega_n} = \dfrac{n\pi v}{2\pi / T_p} = \dfrac{nvT_p}{2} = \dfrac{n\lambda_n}{2}$.

(We remind that T_p is the movement period, and the product, $vT = \lambda$ is the wavelength).

At the end $L = \dfrac{n\lambda_n}{2}$ is called vibration node.

Let's look for the belly of the vibration
In this case the amplitude of the deformation $U(x,t)$ is max.
Using the expression of equation (3), we can note that

$$U(x,t)_{max} \Rightarrow \frac{\omega x}{v} = (2n+1)\frac{\pi}{2}.$$

This gives the following possible values $x_n = (2n+1)\dfrac{\pi v}{2\omega}$.

Taking into account the relationship between the period T_p and the pulsation ω $\left(\omega = \dfrac{2\pi}{T_p}\right)$, we find that $x_n = (2n+1)\dfrac{\pi v}{2\omega} = (2n+1)\dfrac{\pi v T_p}{2 \times 2\pi} = (2n+1)\dfrac{\lambda_n}{4}$.

The vibration bellies are found at the positions: $x_n = (2n+1)\dfrac{\lambda_n}{4}$.

Conclusion
The distance λ_n represents a spatial period $\lambda = 2L$.

The time T_p represents a temporal period $T_p = \dfrac{2L}{v}$.

3. Junction of two vibrating ropes
Consider two (2) infinite ropes joined at a point O.
Rope 1 has linear mass μ_1 and is before the origin point O.
(i.e. its position is defined by $x < 0$).
Rope 2 has linear mass μ_2 and is after point O.
(i.e. its position is defined by $x < 0$).

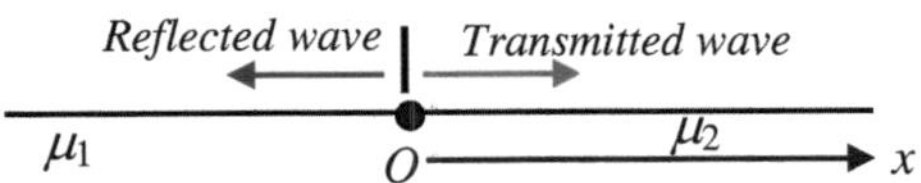

Consider an incident wave (IW) of -∞, when it arrives at the point O (obstacle), a part is transmitted towards the part defined by $x > 0$ (it is called transmitted wave TW) and a part is reflected in the part $x < 0$ (known as reflected wave RW). So, IW = TW + RW.

Transmission and reflection of waves
We consider a strand of the rope centered on O.
It is assumed that the two strings are subjected to the same tension

The deformation in the 1st
rope ($x < 0$) is noted $U_1(x,t)$

The deformation in the 2nd
rope ($x > 0$) is noted $U_2(x,t)$

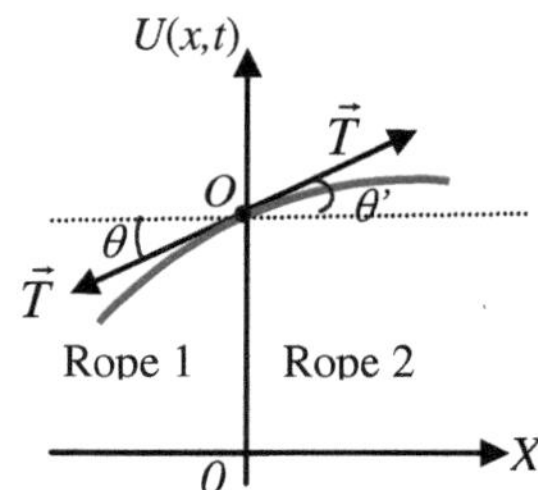

the projection give $T\sin(\theta) = T\sin(\theta')$
$\Rightarrow \sin(\theta) = \sin(\theta')$. Consequently, $\theta = \theta'$ and $\tan(\theta) = \tan(\theta')$.

since: $\tan(\theta) = \dfrac{\partial U_1(x,t)}{\partial x}$ and $\tan(\theta') = \dfrac{\partial U_2(x,t)}{\partial x}$.

Then: $\left.\dfrac{\partial U_1(x,t)}{\partial x}\right|_{x=0} = \left.\dfrac{\partial U_2(x,t)}{\partial x}\right|_{x=0}$ (4)

On another hand,

$$U_1(x,t) = f_i\left(t - \frac{x}{v_1}\right) + f_r\left(t + \frac{x}{v_1}\right) \text{ (incident wave + reflected wave).}$$

$$U_2(x,t) = f_t\left(t - \frac{x}{v_2}\right) \text{ (transmitted wave).}$$

Then $v_1 = \sqrt{\dfrac{T}{\mu_1}}$ and $v_2 = \sqrt{\dfrac{T}{\mu_2}}$ are the propagation velocities of the

transverse wave in chords 1 and 2, respectively.
Substituting U_1 and U_2 in equation (4), calculating the derivative,

we obtain $-\dfrac{1}{v_1} f_i' + \dfrac{1}{v_1} f_r' = -\dfrac{1}{v_2} f_t'$ (5)

$$\left.\dfrac{\partial U_1(x,t)}{\partial t}\right|_{x=0} = \left.\dfrac{\partial U_2(x,t)}{\partial t}\right|_{x=0} \qquad (6)$$

$\Rightarrow f_i' + f_r' = f_t'$ (7) (we have derived U_1 and U_2 with respect to time).

In summary, $\begin{cases} f_i' + f_r' = f_t' \\ \dfrac{1}{v_1}(f_i' - f_r') = \dfrac{1}{v_2}f_t' \end{cases}$ (8)

By integrating equation (8) $\Rightarrow \begin{cases} f_i + f_r = f_t \\ \dfrac{1}{v_1}(f_i - f_r) = \dfrac{1}{v_2}f_t \end{cases}$ (9), (for $x = 0$ and $\forall\ t$).

If we pose: $\begin{cases} f_r = -rf_i \\ f_t = tf_i \end{cases}$ with $\begin{cases} r \ \ \textit{refletion coefficient} \\ t \ \ \textit{transmission coefficient} \end{cases}$.

Our writing is justified by the fact that the waves (reflected and transmitted) are generated by the incident wave (IW = TW + RW).

We replace (f_r and f_t) in the equation (9),

$\begin{cases} 1 - r = t \\ \dfrac{1}{v_1}(1 + r) = \dfrac{1}{v_2}t \end{cases} \Rightarrow \begin{cases} 1 = r + t \\ v_2(1 + r) = v_1 t \end{cases} \Rightarrow \begin{cases} r = \dfrac{v_1 - v_2}{v_1 + v_2} = \dfrac{1 - \sqrt{\dfrac{\mu_2}{\mu_1}}}{1 + \sqrt{\dfrac{\mu_2}{\mu_1}}} \\[4mm] t = \dfrac{2v_2}{v_1 + v_2} = \dfrac{2}{1 + \sqrt{\dfrac{\mu_1}{\mu_2}}} \end{cases}$.

Discussion

- If $\mu_1 = \mu_2 \Rightarrow \begin{cases} r = 0 \\ t = 1 \end{cases}$ (no reflection).

- If $\mu_2 \ll \mu_1 \Rightarrow \begin{cases} t = 0 \\ r = 1 \end{cases}$ (total reflection, no propagation in $x > 0$).

4. Impedance

We call impedance at a given point the ratio of the complex amplitude of the force to the complex amplitude of the velocity of the particle.

It is given by $Z(x) = \dfrac{F_y}{\dot{U}(x,t)}$.

F_y is the projection of the force along the $(y'y)$-axis $\Rightarrow F_y = -T\dfrac{\partial U}{\partial x}$.

Remind

The shape of the 1D-impulsion is $U(x,t) = U_0 \cos(\omega t - kx)$.

In complex notation it is written $U(x,t) = U_0 \exp[\,j(\omega t - kx)]$.

So, F_y is expressed as $F_y = jkTU_0 \exp[\,j(\omega t - kx)]$.

The velocity is $\dot{U}(x,t) = \dfrac{\partial U}{\partial t} = j\omega U_0 \exp[\,j(\omega t - kx)]$.

Taking complex notations into account, in the case of the traveling transverse wave, the characteristic impedance of the string is $Z(x)$, given by

$Z(x) = \mu v = \sqrt{\mu T}$, (the velocity of propagation in the rope $v = \sqrt{\dfrac{T}{\mu}}$).

Chapter 7

Propagation of plane waves; longitudinal waves

1. Propagation of plane waves in fluids
- Longitudinal wave

Longitudinal mechanical waves are called acoustic waves, or compression or pressure waves. Propagation media can be fluids (liquids, gases) or solids (metals).

Acoustic waves are mechanical waves generated by the vibration of systems. In this type of wave, the deformation is in the same direction as the propagation.

Acoustic waves are transmitted by elastic material media, which they can deform to propagate without transport of matter.

1.1. Introduction

We consider a homogeneous fluid constituting a 3D medium, such as a liquid or a gas.

We choose a determined direction (for example (OX)), and compress the fluid by a rigid displacement perpendicular to the direction (OX).

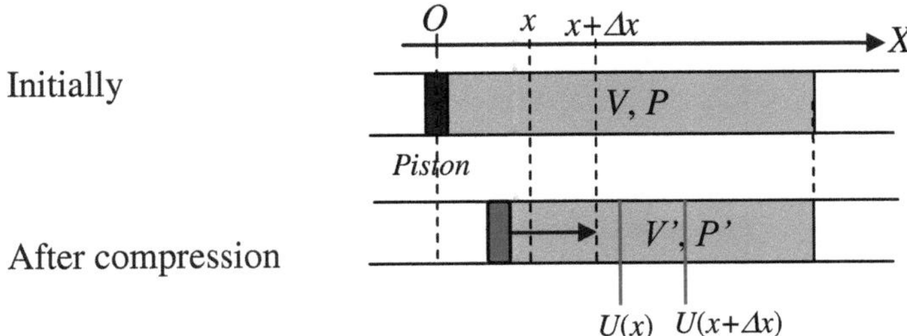

The compression will propagate along the (OX)-axis and the state of the system will be characterized by

- the pressure P,
- the volumic mass ρ,
- the displacement $U(x,t)$.

To have the propagation equation, you must use these three quantities.

Explanation

The disturbance (compression) causes the displacement of the fluid, and this displacement modifies the volumic mass ρ $\Rightarrow$ this inevitably leads to a change in the pressure P.

2. Study of a disturbance in a fluid - Equation

1) Step 1

In the state of equilibrium, the fluid is defined by a pressure P_0, a volumic mass ρ_0.

Out of the equilibrium, the fluid is characterized by a pressure P, a volumic mass ρ that increases, since the fluid has undergone compression.

$$\begin{cases} P = P_0 + \Delta P \\ \rho = \rho_0 + \Delta\rho \end{cases}, \text{ with } \begin{array}{l} \Delta P \prec\prec P_0 \\ \Delta\rho \prec\prec \rho_0 \end{array}.$$

During compression, there is a relationship between P and ρ, and we write $P = f(\rho)$.

$$\Rightarrow P = P_0 + \Delta P = f(\rho_0 + \Delta\rho) = f(\rho_0) + \Delta\rho \times f'(\rho_0).$$

If we pose $\chi = f'(\rho_0) = \left.\dfrac{\partial P}{\partial \rho}\right|_{\rho=\rho_0}$. ($\chi$ is the compressibility coefficient)

So, we can rewrite the expression for pressure as $P = P_0 + \chi\Delta\rho$.

$\Delta P = \chi\Delta\rho$ (1) (this equation is called the equation of state).

Let's find the dimensions of χ:

$$[\chi] = \frac{[\partial P]}{[\partial \rho]} = \frac{ML^{-1}T^{-2}}{ML^{-3}} = L^2 T^{-2} \equiv [v]^2 \ (velocity)^2.$$

2) Step 2

Consider the slice of fluid placed between x et $(x+\Delta x)$ in a pipe

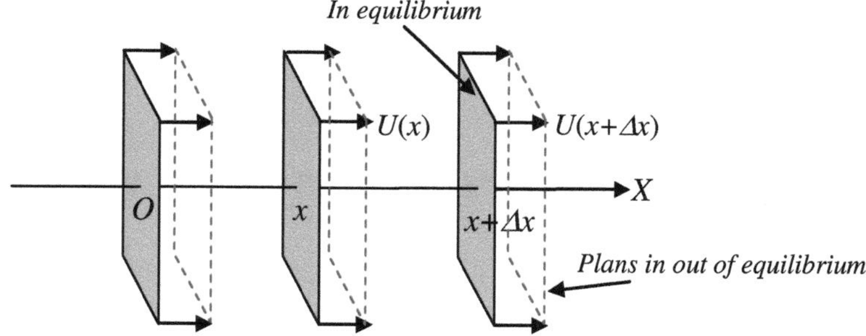

i) The face placed at x (in the state of equilibrium) undergoes a displacement $U(x,t)$ and will be at $(x+U(x,t))$, at the time t.

ii) At the same time, the face positioned at $(x+\Delta x)$ at equilibrium, will be at $[x+\Delta x +U(x+\Delta x,t)]$ (out of equilibrium).

iii) The amount of fluid is constant $\forall\ t$

At the equilibrium   Out of equilibrium

$$S\,\Delta x\,\rho_0 \quad = \quad S\,\rho[x+\Delta x+U(x+\Delta x,t)-x-U(x,t)]$$

Here, S is the pipe section; Δx is the distance between the two planes at equilibrium and $[x+\Delta x+U(x+\Delta x,t)-x-U(x,t)]$ is the distance that separates the two planes after the displacement (out of equilibrium).

$$\Delta x\,\rho_0 = \rho[\Delta x+U(x+\Delta x,t)-U(x,t)]$$

Since $\Delta x \prec\prec$ (very weak) $\Rightarrow U(x+\Delta x,t) = U(x,t) + \Delta x.\dfrac{\partial U(x,t)}{\partial x}$.

Then

$$\rho_0\Delta x = \rho[\Delta x + \Delta x\dfrac{\partial U(x,t)}{\partial x}]$$

$$= (\rho_0 + \Delta\rho)\Delta x[1 + \dfrac{\partial U(x,t)}{\partial x}],$$

$$\Rightarrow \rho_0\dfrac{\partial U(x,t)}{\partial x} + \Delta\rho + \underbrace{\Delta\rho\dfrac{\partial U(x,t)}{\partial x}}_{neglect\ term} = 0 \text{ (After development).}$$

On the other hand, we know that $\Delta\rho \prec\prec \rho_0$, so,

$$\Rightarrow \Delta\rho = -\rho_0\dfrac{\partial U(x,t)}{\partial x} \quad (2) \text{ (called continuity relationship).}$$

Otherwise, $\dfrac{\Delta\rho}{\rho_0} = -\dfrac{\partial U(x,t)S}{\partial x\,S}$ (the relative change in volumic mass is equal

and opposite of the change in volume).

3) Step 3

According to the previous diagram, the section S is $\perp U(x,t)$ (on impulsion). The force acting on the slice of fluid placed at x is

$$F =[P(x,t)-P(x+\Delta x,t)].S$$

Knowing that $\Delta x \prec\prec$ (weak) $\Rightarrow P(x+\Delta x,t) = P(x,t) + \Delta x\dfrac{\partial P(x,t)}{\partial x}$.

As $P = P_0 + \Delta P \Rightarrow P(x + \Delta x, t) = P(x,t) + \Delta x \dfrac{\partial(\Delta P)}{\partial x}$.

Consequently, $F = -\Delta x\, S\, \dfrac{\partial(\Delta P)}{\partial x}$.

On other hand, $F = m.a = \rho_0 \Delta x S \times \dfrac{\partial^2 U(x,t)}{\partial t^2}$.

$\Leftrightarrow \rho_0 \Delta x S \times \dfrac{\partial^2 U(x,t)}{\partial t^2} = -\Delta x\, S \times \dfrac{\partial(\Delta P)}{\partial x}$.

$\rho_0 \dfrac{\partial^2 U(x,t)}{\partial t^2} + \dfrac{\partial(\Delta P)}{\partial x} = 0$ (3) (so-called hydrodynamic relationship)

Recapitulative
$$
\begin{cases}
\Delta P = \chi \Delta \rho & (1) \\[2mm]
\Delta \rho = -\rho_0 \dfrac{\partial U(x,t)}{\partial x} & (2) \\[2mm]
\rho_0 \dfrac{\partial^2 U(x,t)}{\partial t^2} + \dfrac{\partial(\Delta P)}{\partial x} = 0 & (3)
\end{cases}
$$

If we substitute (1) in (3), we get $\rho_0 \dfrac{\partial^2 U(x,t)}{\partial t^2} = -\chi \dfrac{\partial(\Delta \rho)}{\partial x}$ (4).

By injecting (2) into equation (4), we arrive at

$\dfrac{\partial^2 U(x,t)}{\partial x^2} - \dfrac{1}{\chi}\dfrac{\partial^2 U(x,t)}{\partial t^2} = 0$ (5) (it gives the equation of propagation of the

elongation $U(x,t)$ in the fluid with the velocity $v = \sqrt{\chi}$).

Note 1
1) in the same way, we can also show that
$\dfrac{\partial^2(\Delta P)}{\partial x^2} - \dfrac{1}{\chi}\dfrac{\partial^2(\Delta P)}{\partial t^2} = 0$, and $\dfrac{\partial^2(\Delta \rho)}{\partial x^2} - \dfrac{1}{\chi}\dfrac{\partial^2(\Delta \rho)}{\partial t^2} = 0$.

2) the wave in the fluid propagates with the velocity $v_s = \sqrt{\chi} = \sqrt{\dfrac{\Delta P}{\Delta \rho}}$,

Acoustic waves are classified according to their frequencies:
Infrasound $f < 20$ (Hz),
Sound waves 20 (Hz) $< f < 20$ (kHz) $\Rightarrow 17{,}2$ (m) $> \lambda > 1{,}72$ (cm).
Ultrasound $f > 20$ (kHz),
Hypersound $f > 1$ (GHz).
(See the corresponding frequencies).

Note 2

In compressible fluids, the sound wave is a longitudinal elastic wave, which modifies the density of the medium (therefore the pressure).

3. Application to an ideal (perfect) gas

Remember that gases obey the law, $PV^{\gamma} = cst$

(P is the pressure, V the volume and γ is the critical exponent).

By definition, $\rho = \dfrac{m}{V} \Rightarrow V = \dfrac{m}{\rho}$ (V and ρ vary inversely),

$\Rightarrow P = cst \times \rho^{\gamma}$.

If we derive the pressure P with respect to ρ, $\dfrac{dP}{d\rho} = \gamma \times cste \times \rho^{\gamma-1} = \gamma \dfrac{P}{\rho}$.

Then, the velocity of sound in the perfect gas can be written as

$$\chi = v_s^2 = \frac{dP}{d\rho} = \gamma \frac{P}{\rho}.$$

It can be rewritten $v_s^2 = \gamma \dfrac{P}{\rho} \times \dfrac{V}{V} = \dfrac{\gamma(nRT)}{m} = \dfrac{\gamma nRT}{nM_{mol}} = \dfrac{nRT}{M_{mol}}$.

with T is the temperature in (Kelvin) and M_{mol} is is the molar mass and R is the ideal gas constant (in J/(K.mol)), it is given by the relation

$$R = \frac{C_p}{C_V} \equiv \frac{masive\,heat\,at\,constant\,Pressure}{masive\,heat\,at\,constant\,Volume}.$$

Conclusion

The velocity of sound v_s in a perfect gas only depends on *the temperature* and *the nature of the gas* and not on *the pressure* or *density* of the gas.

4. Acoustic impedance

After writing the propagation equation of the longitudinal (acoustic) wave, we will solve it, in a second step. And this within the framework of the acoustic approximation.

Acoustic approximation
It is assumed that the pressure P and the amplitude of the elongation $U(x,t)$ are weak.

The wave propagation equations (for the 3 quantities characterizing the propagation of the wave at the velocity v_s)

$$\frac{\partial^2 U}{\partial t^2} = v_s^2 \frac{\partial^2 U}{\partial x^2} \text{ (5a)}, \quad \frac{\partial^2 (\Delta P)}{\partial t^2} = v_s^2 \frac{\partial^2 (\Delta P)}{\partial x^2} \text{ (5b)}, \quad \frac{\partial^2 (\Delta \rho)}{\partial t^2} = v_s^2 \frac{\partial^2 (\Delta \rho)}{\partial x^2} \text{ (5c)}$$

We point out that the quantities depend simultaneously on x and t ($U(x,t)$, $P(x,t)$, $\rho(x,t)$).

The solution of equations (5a-5c) are of the same form (see the chapter introduction to propagation phenomena).

By analogy, we can write $\dfrac{\partial^2 \overline{\eta}(x,t)}{\partial t^2} = v_s^2 \dfrac{\partial^2 \overline{\eta}(x,t)}{\partial x^2}$ (6)

$\overline{\eta}(x,t)$ being the velocity of the fluid particles at time t and at position x.

The solutions of equation (6) are $\overline{\eta}(x,t) = f(t - \frac{x}{v_s}) + g(t + \frac{x}{v_s})$, and

$\dfrac{\partial \overline{\eta}(x,t)}{\partial t} = f'(t - \frac{x}{v_s}) + g'(t + \frac{x}{v_s})$ (the derivative with respect to time t).

From equation (2), we can write $\Delta \rho = -\rho_0 \dfrac{\partial \overline{\eta}(x,t)}{\partial x} = \dfrac{\Delta P}{v_s^2}$,

$\Rightarrow \Delta P = -\rho_0 v_s^2 \dfrac{\partial \overline{\eta}(x,t)}{\partial x}$. And $\dfrac{\partial \overline{\eta}(x,t)}{\partial x} = -\dfrac{1}{v_s} f'(t - \frac{x}{v_s}) + \dfrac{1}{v_s} g'(t + \frac{x}{v_s})$

(the derivative with respect to x),
We replace P in the previous equation, we will have

$$\Delta P = \rho_0 v_s \left[f'(t - \frac{x}{v_s}) + g'(t + \frac{x}{v_s}) \right].$$

Discussion
i) Case where no reflection ($g = 0$),

$$\Delta P = \rho_0 v_s f'(t - \frac{x}{v_s}) \text{ and } \overline{\eta}(x,t) = f'(t - \frac{x}{v_s}) \Rightarrow \Delta P = \rho_0 v_s \overline{\eta}(x,t).$$

We define the *acoustic impedance* (iterative)

$$Z_0 = \frac{F_0}{\bar{\eta}} = \frac{\Delta P}{S\bar{\eta}} = \frac{\rho_0 v_s \bar{\eta}}{S\bar{\eta}} = \frac{\rho_0 v_s}{S}\,,\ \text{(the expression } S\bar{\eta} \text{ define the flow)}.$$

ii) Case of existence of reflection ($g \neq 0$)
At each point of the fluid, there is a superposition of two reflected and incident waves.

5. Acoustic energy

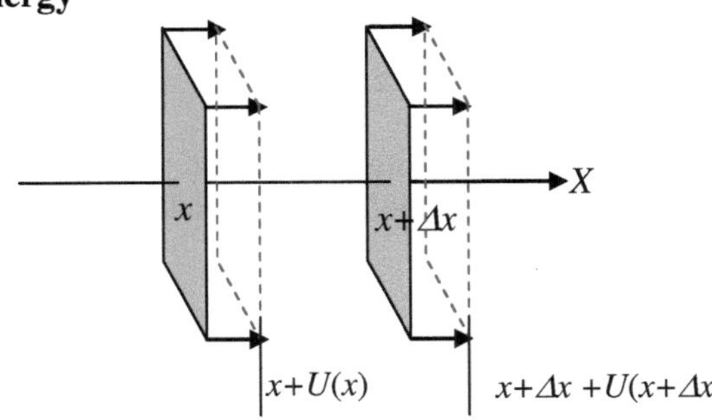

5.1. Kinetic energy

Its expression is $E_c = \dfrac{1}{2} m \left(\dfrac{\partial U}{\partial t} \right)^2 = \dfrac{1}{2} \rho_0 V \left(\dfrac{\partial U}{\partial t} \right)^2$.

Its volume energy density is $\dfrac{E_c}{V} = \dfrac{1}{2} \rho_0 \left(\dfrac{\partial U}{\partial t} \right)^2$.

Its unit is (J/m^3).

5.2. Potential energy

We know: $dE_p = -F.U = -P \underset{volume}{S.U} = -PdV$.

On other hand, $\chi = \dfrac{\partial P}{\partial \rho} = \dfrac{\partial P}{\partial (m/V)} \Rightarrow$ therefore $\partial V = -\chi V \, \partial P$.

At the end, $dE_p = \chi P V \, dP$. By integrating, we obtain $E_p = \dfrac{1}{2} \chi V P^2$.

The volume density of potential energy is $\dfrac{E_p}{V} = \dfrac{1}{2} \chi P^2$.

Its unit is (J/m^3).

5.3. Total energy density

By definition, the energy density W is equal to the energy (kinetic + potential) per unit volume of wave propagation in the fluid.

$$W = \frac{1}{2}\rho_0\left(\frac{\partial U}{\partial t}\right)^2 + \frac{1}{2}\chi P^2 \text{. (Unit of } W \text{ is (J/m}^3\text{))}.$$

6. Propagation of plane waves in a solid bar

All real solid materials are deformable under the action of external stresses. There are two types of deformations:
- Elastic deformations, disappearance of the deformation when the external action ceases.
- Plastic deformations, when the action is removed the deformation remains.
In this course, only the elastic deformations of small amplitudes are studied.

6.1. Propagation equation

We consider a solid bar in which we operate a compression $\Rightarrow$ this generates a deformation $U(x,t)$ (also called displacement of the wave).
Consider a slice between two sections normal to the abscissas (x) and $(x+dx)$.

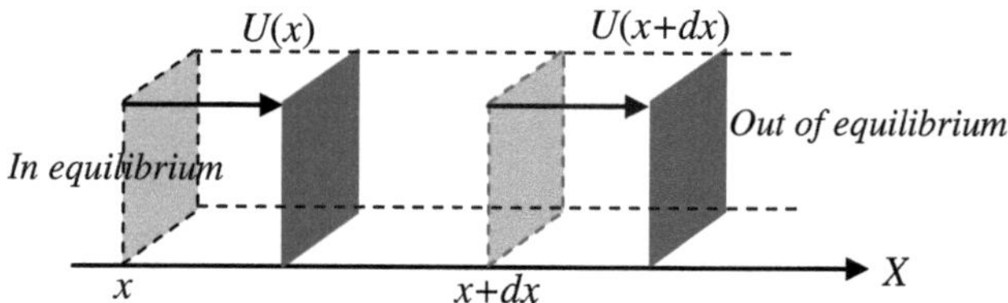

- At rest (at $t = 0$), the length of the slice is dx.
- Out of equilibrium, the length of the slice is

$$[x+dx+U(x+dx)]-[x+U(x)]=dx+\frac{\partial U}{\partial x}dx \text{ .}$$

Following compression, a new length ($\frac{\partial U}{\partial x}dx$) appeared and it is said that

the slice extended by a distance of $\frac{\partial U}{\partial x}dx$.

The relative elongation is given by $\varepsilon = \dfrac{\Delta l}{l} = \dfrac{\partial U}{\partial x}$.

We will have (by application of Newton's law)

$$\rho S\, dx \frac{\partial^2 U}{\partial t^2} = F(x+dx) - F(x) = \Delta F = \frac{\partial F}{\partial x} dx ,$$

Further, $\Delta\rho \lll \rho = \rho_0 \Rightarrow \rho S \dfrac{\partial^2 U}{\partial t^2} = \dfrac{\partial F}{\partial x}$.

In addition, (according to Hooke's law, established experimentally), we know that: $\sigma = \dfrac{F}{S} = \varepsilon E$ (For small deformations).

The symbol σ designates the stresses (or pressure, since it is a force per unit area), the ε is the relative elongation and E is called Young's modulus.

By injecting the expression of the relative elongation ($\varepsilon = \dfrac{\partial U}{\partial x}$), we obtain

$$\frac{F}{S} = \frac{\partial U}{\partial x} E \Rightarrow \frac{\partial U}{\partial x} = \frac{F}{S E} .$$

$$\Rightarrow \frac{\partial^2 U(x,t)}{\partial x^2} = \frac{1}{ES}\frac{\partial F}{\partial x} = \frac{\rho_0}{E}\frac{\partial^2 U(x,t)}{\partial t^2} .$$

At the end, $\dfrac{\partial^2 U(x,t)}{\partial x^2} - \dfrac{\rho_0}{E}\dfrac{\partial^2 U(x,t)}{\partial t^2} = 0$.

The deformation (of the longitudinal wave) propagates in a solid with a velocity $v = \sqrt{\dfrac{E}{\rho_0}}$.

Chapter 8

Electromagnetic waves

1. Electrostatic reminders
1.1. Electric field
During the course of electricity in L1-Level (University first year), we saw that a punctual charge q, isolated and placed at any point O in space, created at another point P (located at a distance r from O) a static electric field $\vec{E}$. It is a radial quantity, and its intensity decreases inversely to the squared distance of r. It is given by the following vector relation

$$\vec{E} = \frac{q}{4\pi\varepsilon_0}\frac{\vec{r}}{r^3} = \frac{q}{4\pi\varepsilon_0 r^2}\vec{U}_{OP},$$ (ε_0 is the vacuum dielectric constant, it is also

called permittivity, $\varepsilon_0 = \dfrac{1}{36\pi\ 10^9}$ (F/m)).

Vector characteristics $\vec{E}$
It essentially depends on the sign of the charge q and the distance r.

If $q > 0$,

If $q < 0$,

Note 1
1) In matter, any charge is a relative number of the elementary charge of the electron, $q = ne$, with $q_e \equiv e = -1.6\ 10^{-19}$ (C).

2) A punctual charge q (placed at point O) creates a uniform electric field around it. This field is decreasing away from O.

3) Any other charge q' placed (in P) in the vicinity of the first charge q undergoes an electric force $\vec{F}_e$ (Coulomb's law) proportional to the charge

q' and the field $\vec{E}$, we write $\vec{F}_e = q'\vec{E} = \dfrac{q'q}{4\pi\varepsilon_0 r^2}\vec{U}_{OP}.$

We note the same phenomenon that the electric charge q' exerts on the charge q, in other words there is a force $-\vec{F}_e$, at the point where the charge q is placed (in O).

4) The electric field created by n punctual electric charges, at the same point, is equal to the sum of the partial fields generated by each of the charges taken separately (principle of superposition), $\vec{E}_{total} = \sum_{i=1}^{n} \vec{E}_i$.

1.2. Electric potential

The space around an electric charge is characterized by two quantities. One is vector (represented by the electric field $\vec{E}$) and the second is scalar, called electrostatic potential $U(r)$.

The relation between the two quantities is given by $\vec{E} = -\overrightarrow{grad}(U(r)$.

Note 2

1) Electrostatic potential is always defined with respect to a reference (like potential energy). Generally, it is considered that the potential is zero when r tends to infinity (far from any influence of charges). We write $U(r \to \infty) = 0$.

2) The partial potentials generated by n punctual electric charges are additive, at a given point in space. The resulting potential is

$$U_{total}(r) = \sum_{i=1}^{n} U_i(r).$$

1.3. Gauss's theorem
1.3.1. Electric field flux

The elementary flow of the electric field $\vec{E}$, through the elementary closed and oriented surface $d\vec{S}$ is a scalar. It is given by $d\Phi = \vec{E}\,d\vec{S}$.

Consequently, the total flow of $\vec{E}$, through the closed surface $\vec{S}$, equals to

$$\Phi = \iint_S d\Phi = \iint_S \vec{E}\,d\vec{S}.$$

1.3.2. Gauss's theorem
i) Punctual electric charges inside a closed surface

If q is the sum of the electric charges lying inside a closed surface $\vec{S}$, it creates, at a distance r, an electric field $\vec{E} = \dfrac{q}{4\pi\varepsilon_0 r^2}\vec{u}$.

The flow of the field $\vec{E}$, created by the charge q, through the closed surface $\vec{S}$, is expressed by

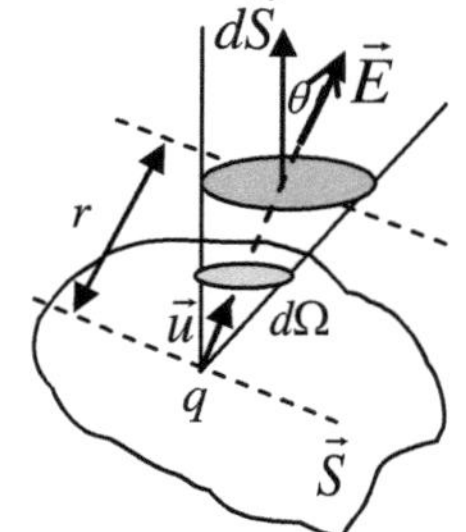

$$\Phi = \iint_S d\Phi = \frac{q}{4\pi\varepsilon_0}\iint_S \frac{\vec{u}\,d\vec{S}}{r^2},$$

$$\Phi = \frac{q}{4\pi\varepsilon_0}\iint_S \frac{dS\cos(\theta)}{r^2} = \frac{q}{4\pi\varepsilon_0}\iint_S d\Omega,$$

$$\Phi = \frac{q}{4\pi\varepsilon_0}(4\pi) = \frac{q}{\varepsilon_0}$$

(the 4π is the solid angle covering the space around the electric charge q).

ii) Punctual electric charges outside a surface

The total flux of the electric field $\vec{E}$ (created by the electric charge q lying outside the surface) through the surface $\vec{S}$ is null.

iii) Case of electric charges distribution on a surface

We examine a set of uniformly distributed electric charges

- Linear electrical charge distribution λ (on a wire)

The linear charge density is given by $\lambda = \dfrac{dq}{dl} \Rightarrow q_L = \int \lambda\, dl$,

- Distribution of electric charges on the surface σ (on a surface)

The surface charge density is $\sigma = \dfrac{dq}{dS} \Rightarrow q_S = \iint_S \sigma\, dS$.

- Volume distribution of electric charges ρ (in a volume)

The volumic charge density is given by $\rho = \dfrac{dq}{dV} \Rightarrow q_V = \iiint_V \rho\, dV$.

- If S designates the surface delimiting the volume V, then the calculation of the flow of the electric field, generated by the homogeneous distribution ρ, is given by $\Phi = \iint_S d\Phi = \iint_S \vec{E}\,d\vec{S} = \dfrac{q_V}{\varepsilon_0} = \dfrac{1}{\varepsilon_0}\iiint_V \rho\, dV$ (this relationship constitutes Gauss's theorem).

- Using the Green-Ostrogradski theorem applied to the electric field $\iint_S \vec{E}\, d\vec{S} = \iiint_V div(\vec{E})\, dV$. We can also write $\iiint_V div(\vec{E})\, dV = \dfrac{1}{\varepsilon_0} \iiint_V \rho\, dV$.

The integration volume is arbitrary and can be taken as small as necessary, we write $div(\vec{E}) = \dfrac{\rho}{\varepsilon_0}$. This relation represents the local form of Gauss's theorem (also known as Coulomb-Gauss).

- Taking into account the relationship between the electric potential and the field, then the local form of Gauss' theorem takes the following form

$\Delta U = -\dfrac{\rho}{\varepsilon_0}$ (Here the symbol Δ is the Laplacian of the electric potential).

The relationship is called the Poisson's equation.

2. Moving charges – Electrokinetics

When electric charges start moving, they generate an electric current. It represents the number of charges crossing a given section per unit time. Its intensity is calculated by $I = \dfrac{dq}{dt}$. Its unit is (C/s $\equiv$ Ampere).

In a conductor, the charge carriers are electrons.

2.1. Electric current density

At each point of the conductive space, two types of densities can be defined: a local electrical charge density and a current density.

Local density of charges, is a scalar that gives the number of charges per unit volume $\rho = \dfrac{dq}{dV}$, its unit is (C/m^3).

Current density is a vector quantity.

At each point M, of a conductive medium, where electric charges move, we define the current density vector $\vec{j}$, having as origin, this same point and for direction and sense that of the movement of the charges.

Its intensity is given by $j = \dfrac{dI}{dS} \equiv \dfrac{I}{S}$, and its unit in (A/m^2).

The dS is the section of a current tube, through which a current intensity dI flows. (Knowing that a current tube is formed by current lines. Each line represents the oriented trajectory of a moving charge).

We note by v and ω, respectively, the velocity and the electric charge density. these notations lead to:

- The charge density is: $\omega = \dfrac{dq}{dV} = \dfrac{dq}{v\,dt\,dS}$,

$\Rightarrow dq = \omega v\,dt\,dS$, (dV is a conductive volume element).

- If replace dq in the expression of dI, we will have

$dI = \dfrac{dq}{dt} = \omega v\,dS \Rightarrow$ the current density, $j = \dfrac{dI}{dS} = \omega v$.

Note 3

The charge density per unit volume is also noted by $\omega = \dfrac{N}{V}$, (N is the

number of charges), then, we can write $j = \left(\dfrac{N}{V}\right) q v = n q v$.

With vectors $\vec{j} = \omega\,\vec{v} = n\,q\,\vec{v}$.

2.2. Charge conservation

The overall charge q contained in an arbitrary conducting volume V, delimited by a closed surface S is given by $q = \iiint\limits_{V} \rho(t)dV$.

When the charge varies over time, the charge variation dq is

$dq = \iiint\limits_{V} \rho(t+dt)dV - \iiint\limits_{V} \rho(t)dV$

$\quad = \iiint\limits_{V} \left[\rho(t+dt) - \rho(t)\right]dV$

$\quad = \iiint\limits_{V} \dfrac{\partial \rho}{\partial t}\,dt\,dV.$

The change in charge dq can be expressed as a function of current density. Let us recall that the variation is due to the displacements of the charges which can penetrate or leave the volume V. If, we note by $d\vec{S}$ the oriented surface element; the charge element dq is then written in the form:

$dq = -\iint\limits_{S} \vec{j}\,d\vec{S}$.

The sign $(-)$ means the charge loss when the charges leave the volume V, ($\vec{j}\,d\vec{S} \succ 0$).

The surface integral can be transformed into a volume integral (Green-Ostrogradski theorem), which give $\iiint_V \left(\dfrac{\partial \rho}{\partial t} + div(\vec{j}) \right) dV = 0$.

Since the considered volume is arbitrary, the integral over this volume can only be zero if the integrand is zero. The charge conservation equation is written in local form $\dfrac{\partial \rho}{\partial t} + div(\vec{j}) = 0$.

3. Magnetic field and magnetostatic

If the electric charge q is moving (e.g., accelerated with an acceleration $\vec{a}(a_T, a_N)$), in addition to the electric field at point P, the charge creates a magnetic induction field $\vec{B}$ *perpendicular* to the electric field $\vec{E}$.

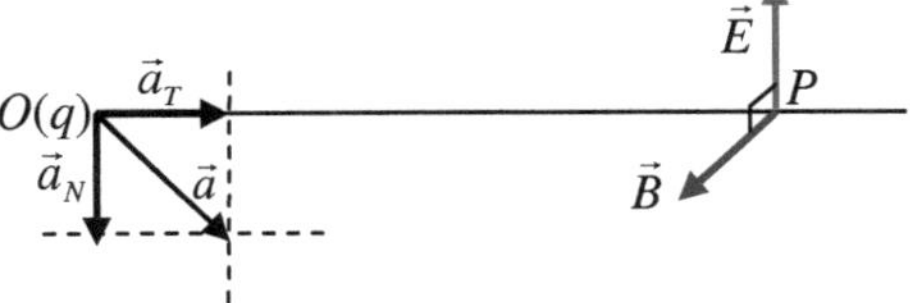

Therefore, the electric current in a conducting wire (due to the mobility of charges) is a source of magnetic field.

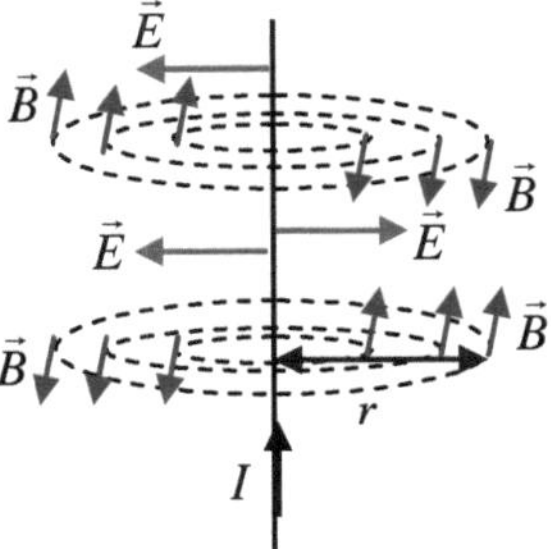

Figure 2a: Magnetic field created by a stationary current in a conducting wire.

- Characteristic of the magnetic field

i) The field $\vec{B}$ is tangent to circular surface of radii r (measurement points). Which gives a field $\vec{B}$ perpendicular to the direction of the current I (Fig. 2a).

ii) The field strength $\vec{B}$, created by a current of intensity I, is $\vec{B} = \dfrac{\mu_0 I}{2\pi r}\vec{e}_\theta$,

(the symbol μ_0 is the magnetic permeability of vacuum $= 4\pi\ 10^{-7}$ (H/m)).

Magnetostatic: studies the magnetic fields created by constant and stationary currents (independent of time).

From the previous examples, we can draw interesting conclusions:

1) Since the charge q is moving $\Rightarrow$ the fields $\vec{E}$ and $\vec{B}$ are propgating.
(If the movement is accelerated, we obtain an electromagnetic wave).

2) The divergence of the field $\vec{B}$ is always null.
(The divergence of the field = the flow of this field through a closed surface based on the Green-Ostrogradski theorem).

We return to the example of current I in the straight wire
We choose a cylindrical surface S (of radius r) whose axis coincides with the wire. Vectors $\vec{B}$ are tangent to S, which gives zero flux.

$\Rightarrow \vec{B}.d\vec{S} = 0$, in any point.

We can write the relation: $\iiint\limits_{V} div(\vec{B})dV = \iint\limits_{S} \vec{B}.d\vec{S} = 0$,

Finally, $div(\vec{B}) = 0$.

3) We have just seen that the $div(\vec{B}) = 0$ always. On the other hand, the field $\vec{E}$ has zero divergence, only in *the absence of charges*. The divergence of an electric field is non-zero in the presence of charges. This is a proof of the inexistence of magnetic monopole (if we break a magnet in two, we do not get two monopoles (north and south) but two magnetic dipole magnets).

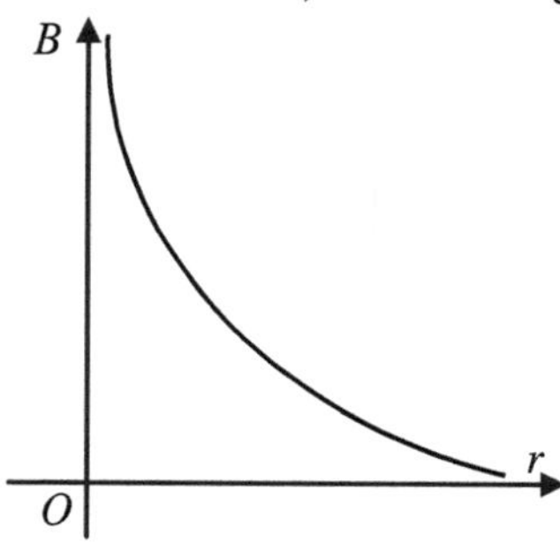

Figure 2b: The plot of the magnetic field created by a straight wire.
(It shows an important rotational character).

4) We now seek the rotational of the field $\vec{B}$

For the calculation, we use the Stokes theorem that transforms the surface integral of the rotational into a contour integral $\oint_C \vec{B}.d\vec{l} = \iint_S \overrightarrow{rot}(\vec{B}).d\vec{S}$,

The S is the surface delimited by the contour C.

By keeping the symmetry of the conductive wire (rectilinear line), we take as contour C a circle perpendicular on the wire, the center of which is located on the conductive wire.

$\Rightarrow$ The contour integral is calculated by $\oint_C \vec{B}.d\vec{l} = \dfrac{\mu_0 I}{2\pi r} 2\pi r = \mu_0 I$.

(We remind that the magnetic field $\vec{B}$ is at a distance r from the wire and on the contour C, moreover it is always collinear with the elementary displacement $\vec{dl}$).

Then, Stokes's theorem gives $\iint_S \overrightarrow{rot}(\vec{B}).d\vec{S} = \mu_0 I$.

Since $\vec{B}$ is oriented in a plane perpendicular to the wire $\Rightarrow$ so, the $\overrightarrow{rot}(\vec{B})$ is parallel to the conducting wire (the current I).

- If, we take a calculation surface S (in the form of a closed cylinder whose axis coincides with the wire), the element $d\vec{S}$ is parallel to the vector $\overrightarrow{rot}(\vec{B})$, only on the circular base of the cylinder.

Also, the current I is calculated from the integral of the current density $\vec{j}$.

$\Rightarrow \iint_S \overrightarrow{rot}(\vec{B}).d\vec{S} = \mu_0 I = \mu_0 \iint_S \vec{j}.d\vec{S}$,

which give $\overrightarrow{rot}(\vec{B})\iint_S d\vec{S} = \mu_0 \vec{j}\iint_S d\vec{S}$.

So, $\overrightarrow{rot}(\vec{B}) = \mu_0 \vec{j}$ (This result constitutes Ampere's theorem).

4. Electromagnetic induction

Electromagnetic induction is the coupling between the two fields $\vec{E}$ and $\vec{B}$. It is obtained, only, in the presence of a variable magnetic field over time.

Faraday's experiment (in 1831)

He varied a magnetic field near a conductor.

Observation an electromotive force capable of moving free electrons is observed (an electric current is generated). He also found that the greater the variation of the magnetic field, the greater the current generated.

Note 4

An identical phenomenon is observed if we use a stationary magnetic field, but with a moving electric circuit.

In the frame of reference of the magnetic field, we denote by the velocity of movement of the free electrons of the circuit. Therefore, these electrons experience the Lorentz force, given by $\vec{F} = q\,\vec{v} \wedge \vec{B} = q\,\vec{E}_m$, with $\vec{E}_m = \vec{v} \wedge \vec{B}$.

Following the observations that arose from the experiments, Faraday proposed a law that relates the flow of the magnetic field ϕ_B to the electromotive force *fem*

$$fem = -\frac{d\phi_B}{dt}.$$

(The *fem* is expressed in Volts, it represents a ddp between two points A and B of the circuit which generates the current I).

The potential difference can also be calculated by the relation

$$fem = \oint_C \vec{E}_m.d\vec{l} \ .$$

Since the divergence of a rotational is always zero $\Rightarrow \dfrac{\partial \rho}{\partial t} = 0$ and the scope of Ampère's theorem is limited to steady states.

In 1876, Maxwell became interested in the analogies between $\vec{E}$ and $\vec{B}$. He noted that only the relation of induction comprises an unsteady term relating to the electric field $\vec{E}$.

Consequently, Maxwell proposed to modify Ampère's theorem to add an unsteady term relating to the magnetic field $\vec{B}$.

The modified relation is written in the form $\overrightarrow{rot}(\vec{B}) = \mu_0\vec{j} - \alpha\dfrac{\partial \vec{E}}{\partial t}$, (with a constant α, to be determined by the conservation of charge).

By applying the divergence to the previous relation, we obtain

$$div\left(\overrightarrow{rot}(\vec{B})\right) = \mu_0 div(\vec{j}) + \alpha\frac{\partial}{\partial t}\left(div(\vec{E})\right)$$

Since the divergence of a rotational is zero, and taking into account the conservation of charge and the local form of Gauss's theorem

$$0 = -\mu_0 \frac{\partial \rho}{\partial t} + \frac{\alpha}{\varepsilon_0} \frac{\partial \rho}{\partial t} \Rightarrow \alpha = \varepsilon_0 \mu_0 \,.$$

The principle of conservation of charge becomes compatible with the *Maxwell-Ampere* relation, written in the form

$$\overrightarrow{rot}(\vec{B}) = \mu_0 \left(\vec{j} + \varepsilon_0 \frac{\partial \vec{E}}{\partial t} \right) = \vec{j} + \vec{j}_D \,.$$

The term $\vec{j}_D$ is added by Maxwell. It comes in the form of a current (known as displacement current).

Note 5

Maxwell never saw experimental verification of his displacement current hypothesis. It was only in 1887 with the work of Hertz that Maxwell's theory was verified.

In summary, Maxwell's four equations are
$$\begin{cases} div(\vec{E}) = \dfrac{\rho}{\varepsilon_0}, \\[2mm] div(\vec{B}) = 0, \\[2mm] \overrightarrow{rot}(\vec{E}) = -\dfrac{\partial \vec{B}}{\partial t}, \\[2mm] \overrightarrow{rot}(\vec{B}) = \mu_0 \left(\vec{j} + \varepsilon_0 \dfrac{\partial \vec{E}}{\partial t} \right). \end{cases}$$

These equations show the coupling between electric and magnetic fields, involving the charge density ρ and the current density $\vec{j}$ as well as two constants ε_0 and μ_0.

5. Electromagnetic waves

We saw that in the previous part that

The propagation of $\vec{E}$ and $\vec{B}$ from the charge q in accelerated motion generates an electromagnetic wave with a vector character.

5.1. Definition of an electromagnetic wave (EMW)

An electromagnetic wave (EMW) is the phenomenon resulting from the simultaneous propagation in space of an electric field and a magnetic induction field with the same velocity and carrying radiant energy.

Like the other waves, we will establish the propagation equation of the EMW.

5.2. Propagation equation

From Maxwell's equations, we can obtain the equation of propagation of the electromagnetic wave.

Indeed, the rotational of the 3rd equation (also known as the Maxwell-Faraday relation) gives $\overrightarrow{rot}\left(\overrightarrow{rot}(\vec{E})\right) = -\overrightarrow{rot}\left(\dfrac{\partial \vec{B}}{\partial t}\right)$.

We note that the variables x, y, z (in space) and time t are independent, so we can permute the time derivative with the rotational operator.

On the other hand, we know that, for a vector $\vec{E}$

$$\overrightarrow{rot}\left(\overrightarrow{rot}(\vec{E})\right) = \overrightarrow{grad}(div(\vec{E}) - \Delta\vec{E}$$

Therefore: $\overrightarrow{grad}\left(div(\vec{E})\right) = -\Delta\vec{E} = -\dfrac{\partial}{\partial t}\left(\overrightarrow{rot}(\vec{B})\right)$

Using the 1st relation of Maxwell's equations, (the divergence of $\vec{E}$) and the 4th relation (known as Maxwell-Ampère), we obtain *the propagation equation of the electric field*

$$\Delta\vec{E} - \dfrac{1}{c^2}\dfrac{\partial^2 \vec{E}}{\partial t^2} = \mu_0 \dfrac{\partial \vec{j}}{\partial t} + \dfrac{1}{\varepsilon_0}\overrightarrow{grad}(\rho). \text{ With a speed } c = \dfrac{1}{\sqrt{\varepsilon_0 \mu_0}}.$$

Likewise, the rotational of Maxwell's 4th relation gives

$$\overrightarrow{rot}\left(\overrightarrow{rot}(\vec{B})\right) = \mu_0 \overrightarrow{rot}(\vec{j}) + \varepsilon_0 \mu_0\left(\dfrac{\partial \vec{E}}{\partial t}\right),$$

By combining it with the 2nd Maxwell relation (divergence of $\vec{B}$)

$\Rightarrow$ the 3rd relation gives the *equation of propagation of the magnetic field*

$$\Delta\vec{B} - \dfrac{1}{c^2}\dfrac{\partial^2 \vec{B}}{\partial t^2} = -\mu_0 \overrightarrow{rot}(\vec{j}). \text{ With a speed } c = \dfrac{1}{\sqrt{\varepsilon_0 \mu_0}}.$$

The electric and magnetic fields therefore propagate at the same speed, called *the speed of light.*

5.3. Light – Wave pattern

Light is an electromagnetic wave which propagates in vacuum (the quantities which propagate are the two fields $\vec{E}$ and $\vec{B}$).

Such a wave is characterized by:

- the phenomena of reflection, refraction, interference and diffraction,

- its speed (celerity) of propagation $c = 3 \times 10^{8}$ (m/s), in the void,
- its frequency f,
- its wavelength λ (in the vacuum).

These quantities are linked by the relations $\lambda = cT = \dfrac{c}{v}$.

The classification of EM waves, into different domains, is done according to their spectra.

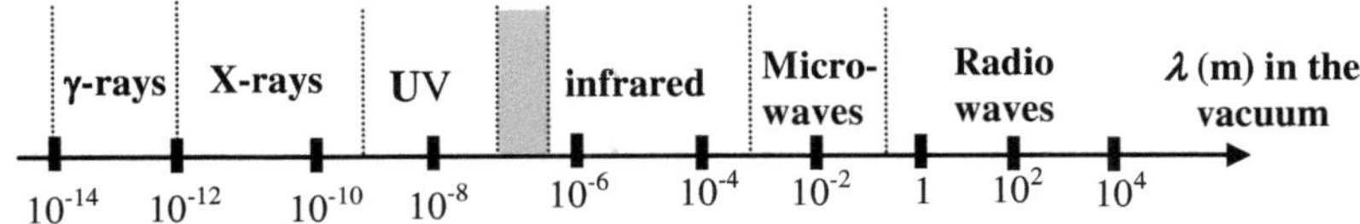

Visible light is the area to which the eye is sensitive. This domain (grayed out above) is such that 400 (nm) $< \lambda <$ 800 (nm).

- Propagation of light in transparent media
The refractive index (noted n) of a medium is the ratio of the celerity c (of

light in a vacuum) to the propagation speed v of light in this medium $n = \dfrac{c}{v}$.

The refractive index depends on the transparent medium considered, (n_{water} = 1.33; n_{glass} = 1.5; n_{air} = 1.00).

Note 6
- The refractive index also depends on the frequency of light.
- The frequency of monochromatic radiation does not change when it passes from one transparent medium to another (otherwise its color would change).
- All transparent media (other than air) are more or less dispersive.

5.3. Propagation of EMW in a vacuum
In the absence of charges and current, the densities $\rho = 0$ and $\vec{j} = 0$,

Maxwell's 4 equations (in vacuum) are
$$\begin{cases} div(\vec{E}) = 0, \\ div(\vec{B}) = 0, \\ \overrightarrow{rot}(\vec{E}) = -\dfrac{\partial \vec{B}}{\partial t}, \\ \overrightarrow{rot}(\vec{B}) = \mu_0 \varepsilon_0 \dfrac{\partial \vec{E}}{\partial t}. \end{cases}$$

The propagation of the electric field $\vec{E}$ reduces to $\left(\Delta - \dfrac{1}{c^2} \dfrac{\partial^2}{\partial t^2} \right) \vec{E} = 0$.

The same work with the field $\vec{B}$ gives $\left(\Delta - \dfrac{1}{c^2} \dfrac{\partial^2}{\partial t^2} \right) \vec{B} = 0$.

Proprieties
- We realize that the propagation of the two fields occurs at the same speed. c).

- Integration in relation to $\phi \Rightarrow$ $\begin{cases} E_z = -c\,B_y \\ E_y = c\,B_z \end{cases}$ and $|\vec{E}| = c|\vec{B}|$.

- The dot product $\vec{E}.\vec{B} = E_y B_y + E_z B_z = c\left[B_y B_z - B_y B_z \right] = 0$, ($\vec{E} \perp \vec{B}$).

The plot of the plane wave EM in the case where the vectors $\vec{E}$ and $\vec{B}$ keep the same direction during the propagation, is given in the figure 3.

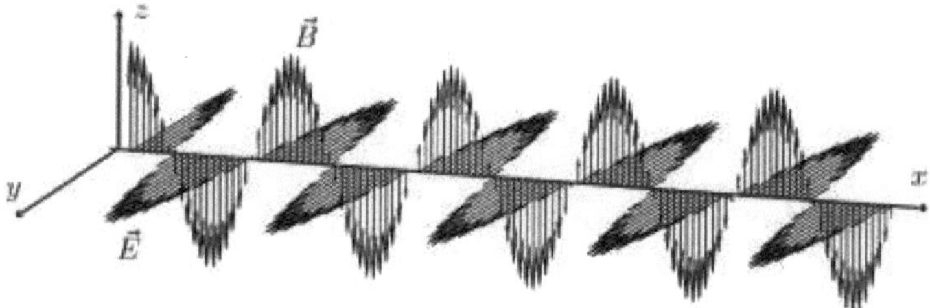

Figure 3: The fields $\vec{E}$ and $\vec{B}$ for a plane wave with rectilinear polarization. (The propagation is done along the x-axis, the field $\vec{E}$ is oriented along the y-axis and and the field $\vec{B}$ is oriented along z-axis).

The monochromatic wave formalism allows the fields to be written in the form $\vec{E} = \vec{E}_0 e^{j(\vec{k}.\vec{r}-\omega t)}$ and $\vec{B} = \vec{B}_0 e^{j(\vec{k}.\vec{r}-\omega t)}$.

The calculation of the divergence relations of the two fields gives

$\vec{k}.\vec{E} = 0$ and $\vec{k}.\vec{B} = 0$.

(This demonstrates the transverse character of the plane wave in vacuum).

Since the two fields are perpendicular, the induction relation imposes $j\vec{k} \wedge \vec{E} = j\omega \vec{B}$. Either $\vec{B} = \dfrac{1}{\omega}\vec{k} \wedge \vec{E} = \dfrac{1}{c}\vec{\kappa} \wedge \vec{E}$.

5.4. Polarization of electromagnetic waves

The vectors $\vec{E}$ and $\vec{B}$ are always perpendicular to the direction of propagation. But in the plan, they can move and describe different configurations. The direction and trajectory described by the electromagnetic fields during propagation is called *the polarization* of the wave.

Polarization

The wave vector orientation $\vec{k}$ relative to the direction of travel of $\vec{E}$ (or of $\vec{B}$).

- Different types of polarization

1) Longitudinal polarization, the wave vector $\vec{k}$ is parallel to the oscillations direction of the field $\vec{E}$ (or $\vec{B}$).

2) Transversal polarization, the wave vector $\vec{k}$ is perpendicular to the oscillations direction of the field $\vec{E}$ (or $\vec{B}$), we distinguish:

*i) **Linear polarization***, the elongation vector of the electromagnetic field, located in a plane perpendicular to the direction of propagation $\vec{k}$, keep a fixed direction in this plane.

*ii) **Elliptical polarization***, the vector $\vec{E}$ (or $\vec{B}$) rotates in the plane perpendicular to the direction of propagation $\vec{k}$. The end of the vector $\vec{E}$ (or $\vec{B}$) describes an ellipse in the plane perpendicular to $\vec{k}$.

Example of representation

Figure 4: Circular polarization of the electric field vector.

(a) The end of the field $\vec{E}$ describes a propeller in space.
Looking at the wave from a point $x \to \infty$, the end of the vector is subject to a circle. The direction of travel of this circle gives: (b) the direction of the circular polarization at the top, (c) circular polarization at the bottom).

In the case of a monochromatic plane wave propagating in the x-direction, the components of the electric field vector are:

$$\vec{E}\begin{pmatrix} 0 \\ E_{y0}\cos(k\,x - \omega t + \phi_y) \\ E_{z0}\cos(k\,x - \omega t + \phi_z) \end{pmatrix}$$

At a fixed position x: the end of the vector $\vec{E}$ describes different figures depending on the value of the dephasage $\phi_y - \phi_z$ between the components y and z.

Discussion of certain situations

1) If $\phi_y - \phi_z$ is a multiple of π, the vector $\vec{E}$ keeps a constant direction. The polarization is rectilinear.

2) If $\phi_y - \phi_z = \pm\dfrac{\pi}{2}$ and $E_{y0} = E_{z0}$, the polarization is circular. The sign of the dephasage determines the direction of rotation of the vector $\vec{E}$ around the x-axis.

For an observer placed at $x = \infty$, (he receives the EMW), the polarization is right circular if the rotation is clockwise. In the opposite case, the polarization is said to be left circular.

3) If $\phi_y - \phi_z = \pm\dfrac{\pi}{2}$ and $E_{y0} \neq E_{z0}$, the end of the vector $\vec{E}$ describes an ellipse, with a direction of circulation given by the dephasage. We rather speak of right or left elliptical polarization, (with the same convention as for circular polarization).

4) If $\phi_y - \phi_z$ is arbitrary and $E_{y0} \neq E_{z0}$, the polarization is elliptical, but the main axes of the ellipse do not coincide with the y and z axes of the coordinate frame.

5.5. Electromagnetic energy

Besides the conservation of charge, Maxwell's equations must verify the principle of conservation of energy. The Maxwell-Ampere relation gives the current density in the form $\vec{j} = \varepsilon_0 c^2 \left[\overrightarrow{rot}(\vec{B}) - \dfrac{1}{c^2} \dfrac{\partial \vec{E}}{\partial t} \right]$.

The dot product with the electric field is $\vec{j}.\vec{E} = \varepsilon_0 c^2 \left[\overrightarrow{rot}(\vec{B}).\vec{E} - \dfrac{1}{2c^2} \dfrac{\partial E^2}{\partial t} \right]$.

(The multiple product in the hook can be subdivided as follows
$\left(\overrightarrow{rot}(\vec{B}) \right)\vec{E} = \left(\overrightarrow{rot}(\vec{E}) \right)\vec{B} + div(\vec{B} \wedge \vec{E})$).

By applying the Maxwell-Faraday relationship $rot(\vec{E}) = -\dfrac{\partial \vec{B}}{\partial t}$, we get a conservation relation: $\vec{j}.\vec{E} + \dfrac{\partial}{\partial t}\left[\underbrace{\dfrac{\varepsilon_0}{2}(E^2 + c^2 B^2)}_{U} \right] + div\left[\underbrace{\varepsilon_0 c^2 (\vec{E} \wedge \vec{B})}_{\vec{\Pi}} \right] = 0$.

In the expression, we note two quantities (scalar U and vector $\vec{\Pi}$)

i) Scalar, called electromagnetic energy density

$$U = \dfrac{\varepsilon_0}{2}(E^2 + c^2 B^2) \text{, (measured in (J/m}^3\text{))}.$$

ii) Vector: called Poynting vector (designates the direction of propagation of energy), $\vec{\Pi} = \varepsilon_0 c^2 (\vec{E} \wedge \vec{B}) = \dfrac{1}{\mu_0} \vec{E} \wedge \vec{B}$.

In the vacuum

We know that $\rho = 0$ and $\vec{j} = 0$. The calculation of the two previous quantities gives

- Poynting's vector, $\vec{\Pi} = \dfrac{1}{\mu_0}\vec{E} \wedge \vec{B} = \dfrac{1}{\omega\mu_0}\vec{E} \wedge (\vec{k} \wedge \vec{E}) = \dfrac{E^2}{\omega\mu_0}\vec{k} = \dfrac{E^2}{c\mu_0}\vec{\kappa}$.

- The energy density of the wave, $U = \dfrac{\varepsilon_0}{2}(E^2 + c^2 B^2) = \dfrac{\varepsilon_0 E^2}{2} + \dfrac{\varepsilon_0 E^2}{2} = \varepsilon_0 E^2$,

(electric and magnetic fields contribute equally to the electromagnetic energy density).

Exercise series n° 3 (Waves)

Exercise 3.1

1) Which of the following expressions satisfies the wave equation?

a) $\psi_1(x,t) = Ae^{-(2kx+3\omega t)^2}$;

b) $\psi_2(x,t) = A\sin(\chi x^2 - \omega t)$;

c) $\psi_3(x,t) = A\sin(kx - \omega t) + A\cos(kx - \omega t)$;

d) $\psi_4(x,t) = A\sin(kx)B\cos(\omega t)$.

2) Simply define traveling wave, regressive wave and standing wave.

Exercise 3.2

A point simultaneously performs two oscillations in the same direction expressed by $x_1(t) = a\cos(\omega t)$ and $x_2(t) = a\cos(2\omega t)$.

- Determine the velocity maximum of this point if: $a = 50$ (mm) and $\omega = 10^3$ (rd/s)

Exercise 3.3

Functions $F(x\text{-}vt)$ and $G(x+vt)$ are defined by

$$F(x) = \begin{cases} 1, & for\ r - 4 \prec x \prec -2 \\ 0 & elsewhere \end{cases}, \quad G(x) = \begin{cases} -1, & for\ 2 \prec x \prec 4 \\ 0 & elsewhere \end{cases}.$$

- Plot the resulting wave $F+G$ for $v = 3$, at times $t = 0$; 1 and 2 (s).

Exercise 3.4

Let consider the progressive plane wave be defined by the function:
$\psi(x,t) = \psi_0 \sin[2\pi(50t - 2x)]$.

1) Calculate its period, its frequency, the associated wavelength and its phase velocity.

2) Give the geometric form of the phase at a time $t > 0$. Make a representation.

3) Evaluate Δx for $\Delta\phi = \dfrac{\pi}{3}$.

4) Evaluate $\Delta\varphi$ for $\Delta t = 1$ (ms).

Exercise 3.5

We give the following two wave functions

$\psi_1(x,t) = A_1 \cos(\omega t - \vec{k}.\vec{r})$ and $\psi_2(x,t) = A_2 \cos(\omega t - k.r)$.

- Compare the phase surfaces of the two functions.

Exercise 3.6

1) Determine the dispersion relationships for the following wave equations

i) $a\dfrac{\partial^2\psi}{\partial t^2} - b\dfrac{\partial^2\psi}{\partial x^2} = 0$;

ii) $\dfrac{\partial^2\psi}{\partial x^2} + \alpha\dfrac{\partial^4\psi}{\partial x^4} - \dfrac{1}{c^2}\dfrac{\partial^2\psi}{\partial t^2} = 0$;

iii) $\dfrac{\partial^2\psi}{\partial z^2} - \beta\dfrac{\partial\psi}{\partial t} - \dfrac{1}{c^2}\dfrac{\partial^2\psi}{\partial t^2} = 0$.

2) Specify for each case whether the propagation is dispersive or not.

Exercise 3.7

- Calculate the velocity of propagation of transverse waves for a steel piano string, assumed to be non-rigid and undamped. The tension force is 900 (N), and the string has a diameter of 3 (mm). we give $\rho = 7900$ (kg/m^3).

Exercise 3.8

A string having the length $L = 2$ (m) and the mass $m = 10$ (g) is stretched between two fixed points with a tension $T = 10$ (N).
1) Calculate the own frequencies of transverse vibrations.
2) Draw the string as it oscillates in the fundamental mode and in the first three modes.

Exercise 3.9

An aluminum wire of length $L_1 = 60$ (cm) whose cross-sectional area is $S_1 = 10^{-2}$ (cm^2) is attached to another iron wire of the same section and length $L_2 = 86.6$ (cm). At one of the two ends of the mixed wire is attached a mass $m = 10$ (kg), arranged as indicated in the figure. We take $g = 9.81$ (m/s^2).

$\rho_{Al} = 2.60$ (g/cm^3)
$\rho_{Fe} = 7.80$ (g/cm^3)

1) Find the own frequency of the transverse vibration, the lowest and for which the junction of the wires is a node.
2) What is the total number of knots that can be observed at this frequency, if those at the ends of the wire are not counted.

Exercise 3.10

A rope of length L and linear mass μ is is stretched by the tension T_0. Both ends being fixed.
1) Determine the wave propagation equation.
2) We are looking for solutions of the form $y(x,t) = F(x)f(t)$ (separate form), determine the general solution of the propagation equation and the vibration frequencies.
3) The rope is now moved away from its equilibrium position at the moment $t = 0$, then released without initial velocity. Give in this case the general propagation equation. We take $y(x,0) = \dfrac{2h}{L}x$ for $0 \leq x \leq L/2$, and

$$y(x,0) = \frac{2h}{L}(L - x) \text{ for } L/2 \leq x \leq L.$$

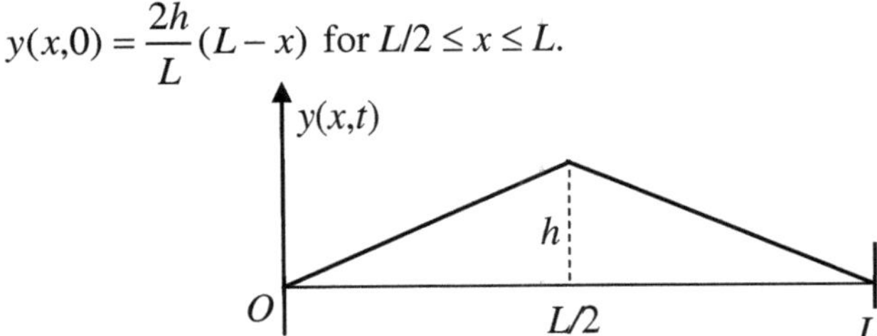

Exercise 3.11

Consider a disturbance that propagates without deformation in a non-dispersive medium along an axis (Ox).
Let note $\Psi(t)$ be the law of variation of this disturbance at the origin O. If it propagates at velocity v, it reaches the point M with positive abscissa x after a time $\tau = x/v$. The vibration law in M is then $\psi(t - \tau) = \psi(t - \dfrac{x}{v})$.

1) By noting $u = t - \dfrac{x}{v}$, show that $\dfrac{\partial^2 \psi}{\partial x^2} - \dfrac{1}{v^2}\dfrac{\partial^2 \psi}{\partial t^2} = 0$.

2) We seek to see if there exist solutions of the form $\psi = F(x)f(t)$.

- Demonstrate that $v^2 \dfrac{1}{F(x)}\dfrac{\partial^2 F(x)}{\partial x^2} = \dfrac{1}{f(t)}\dfrac{\partial^2 f(t)}{\partial t^2}$.

This equation only admits a solution if its two sides are equal to a constant. Why? We set this constant equal to $-\omega^2$; calculate $F(x)$ and $f(t)$, (we pose $k = \dfrac{\omega}{v}$).

Exercise 3.12

A transverse sine wave of frequency v and of amplitude a propagates along a stretched string parallel to the (Ox)-axis, with a celerity c in the direction of decreasing x.

Data: $f = 6$ (Hz), $c = 24$ (m/s) and $a = 3$ (mm).

1) Write the expression for the displacement $\Psi(x,t)$ from one point of the rope knowing that $\Psi(0,0) = 0$. We will show that there exist two solutions.

2) Check that the abscissa point $+ 0.5$ (m) has a positive displacement at the moment $t = 0$.

3) Give the expression for the velocity $v(x,t)$ of one point on the rope.

Exercise 3.13

Two elongations $S_1(x,t) = S_0 \cos(\omega t - kx - \varphi_1)$ and $S_2(x,t) = S_0 \cos(\omega t - kx - \varphi_2)$ propagate along a rope.

1) Give the direction of propagation of each of the excitations.

2) Give the expression for the resulting wave $S(x,t) = S_1(x,t) + S_2(x,t)$.

   a) Determine the vibration points for which $S(x,t) = 0$.

   - Deduce the distance between two consecutive points.

   b) Determine the vibration points for which $|S(x,t)|$ is maximum.

Exercise 3.14

Consider a vibrating rope of length l, of linear mass μ and subjected to a tension T_0.

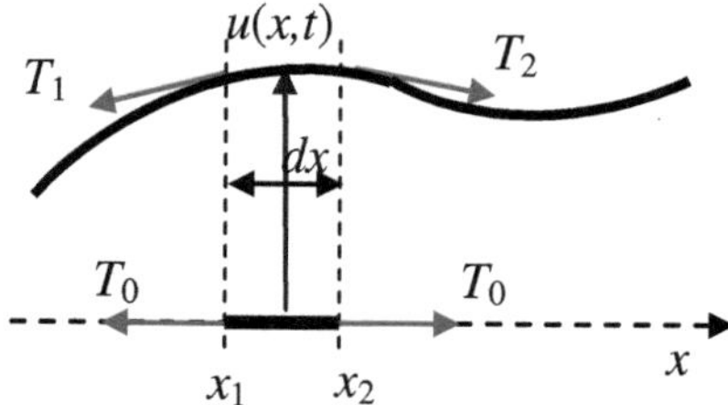

1) The rope is fixed at its ends in $x = 0$ and in $x = l$.

- Give the differential equation that governs transverse displacements $u(x,t)$ of the rope as a function of its characteristics.

2) Demonstrate that the stationary solution of the propagation equation can be put, for a mode of vibration, in the form $u(x,t) = f(x)\cos(\omega t + \varphi)$

- Determine the expression of $f(x)$.

3) Taking into account that the rope is fixed in $x =$ and $x = l$, then express $u(x,t)$ for one mode.

4) Deduce in the form of a Fourier series the general solution of the movement of this rope.

5) Knowing that at the moment $t = 0$, $\dot{u}(x,t) = 0$, how to simplify this expression?

6) The shape of the rope at $t = 0$ being: $f(x) = a\sin(\dfrac{3\pi}{l}x)$,

- what is then the general solution of the movement of this rope?

Exercise 3.15

The oscillator resulting from the superposition of two harmonic oscillations of the same direction has the expression $x(t) = A\cos(2.1t)\cos(50t)$.

- Determine the pulsations (or frequencies) of the component oscillations and the beat period of the resulting oscillation.

Exercise 3.16

Any wave state can always be in the form of a discrete or continuous series of sinusoidal functions (wave front or wave train).

1) Define the phase and group velocities. Deduce the relationship linking the two velocities (Rayleigh criterion). Study the case where $v_\varphi = \dfrac{c}{n}$

2) Calculate the ratio $\dfrac{v_\varphi}{v_g}$ for $\dfrac{dn}{d\lambda} = -2\ 10^5\,(\text{m})$, $\lambda = 0.55\ (\mu\text{m})$ and $n = 1.64$ (refractive index).

3) We consider for a given wave $f^2 = c^2\left(\dfrac{k^2}{4\pi^2} + \dfrac{b^2}{4a^2}\right)$, where f is the frequency, k is the value of the wave-vector, b is an integer and a is a constant.

    $i)$ Calculate v_φ as a function of a, b, c and k.

    $ii)$ Demonstrate that $v_\varphi \times v_g = c^2$, and deduce v_g.

Are the requirements of relativistic mechanics contradicted?
iii) Repeat question *i*) in the case of a non-dispersive medium.

Exercise 3.17

An electromagnetic plane wave has an electric field $\vec{E} = \left[E_0 e^{j(\vec{k}.\vec{r}-\omega t)} \right] \vec{e}_x$.

- Specify the orientation of the wave vector and determine the corresponding magnetic field.

Exercise 3.18

In a planar waveguide, the electric field is described by

$$\vec{E} = E_0 \left[\sin(\frac{n\pi y}{L_y}) e^{j(k_g x - \omega t)} \right] \vec{e}_z .$$

- Calculate the components of the magnetic field $\vec{B}$ associated with him.

Exercise 3.19

During the propagation of a 1D acoustic wave, in a fluid, the pressure is described by a function $P(x,t) = P_0 e^{j(\omega t - kx)}$.

And the dispersion relation is $D(k,\omega) = \omega^2 - v_a^2 k^2 (1 + j\omega_\tau) = 0$.

Where v_a is the velocity of the wave without attenuation and τ is a characteristic time which depends on the viscosity η of the liquid and the volumic mass ρ_0.

It is given by the relationship $\tau = \dfrac{\eta}{\rho_0 v_a^2}$.

1) Calculate the order of magnitude of the quantity ω_τ for an ultrasonic wave of frequency 1 (MHz). We take $v = 1500$ (m/s) and $\eta = 10^{-3}$ (Pa.s).
2) By justified approximation, write the relation in the form of:
  $k = k_{re}(\omega) + j \times k_{im}(\omega)$.
3) Deduce that the pressure can be written in the form

$P(x,t) = P_0 e^{-x/L} e^{j(\omega t - k_{re}x)}$, where L is a characteristic attenuation distance.

Exercise 3.20

Two electromagnetic plane waves of pulsation ω propagate according to the wave vectors $\vec{k}_1$ and $\vec{k}_2$, such as

$\vec{k}_1 = \cos(\theta)\vec{e}_y + \sin(\theta)\vec{e}_z$ and $\vec{k}_2 = \cos(\theta)\vec{e}_y - \sin(\theta)\vec{e}_z$.

The electric fields of these two waves are

$$\vec{E}_1 = E_0 e^{j(\vec{k}_1 \cdot \vec{r} - \omega t)} \vec{e}_x \text{ and } \vec{E}_2 = E_0 e^{j(\vec{k}_2 \cdot \vec{r} - \omega t)} \vec{e}_x .$$

1) Calculate the corresponding magnetic fields $\vec{B}_1$ and $\vec{B}_2$.

2) Calculate the real parts of the electric and magnetic fields resulting from the superposition of the two waves.

3) Calculate the Poynting vector corresponding to the superposition of the two waves, and its temporal average.

4) Show that electromagnetic energy presents, on a time average, an interference structure in the plane $y = 0$.

- Give the distance between two successive interference fringes.

Answer key to exercise series n° 3

Exercise 3.1

The equation of wave is given by $\dfrac{\partial^2\psi}{\partial x^2}-\dfrac{1}{v^2}\dfrac{\partial^2\psi}{\partial t^2}=0 \Rightarrow \dfrac{\partial^2\psi}{\partial t^2}\Big/\dfrac{\partial^2\psi}{\partial x^2}=v^2$.

a) $\dfrac{\partial^2\psi_1}{\partial x^2}=2k(2kx+3\omega t)\psi_1$ and $\dfrac{\partial^2\psi_1}{\partial t^2}=3\omega(2kx+3\omega t)\psi_1$.

The report $\dfrac{\partial^2\psi_2}{\partial t^2}\Big/\dfrac{\partial^2\psi_2}{\partial x^2}=\dfrac{2k(2kx+3\omega t)\psi_1}{3\omega(2kx+3\omega t)\psi_1}$ does not represent a speed;

therefore, the function Ψ_1 does not verify the wave equation.

b) $\dfrac{\partial^2\psi_2}{\partial x^2}=2\chi\,x\psi_2$ and $\dfrac{\partial^2\psi_2}{\partial t^2}=-\omega\psi_2$.

The report $\dfrac{\partial^2\psi_2}{\partial t^2}\Big/\dfrac{\partial^2\psi_2}{\partial x^2}=\dfrac{2\chi\,x}{-\omega}$ does not represent a speed, as consequence

the function Ψ_2 does not verify the wave equation.

c) $\dfrac{\partial^2\psi_3}{\partial x^2}=Ak\big[\cos(kx-\omega t)-\sin(kx-\omega t)\big]$

and $\dfrac{\partial^2\psi_3}{\partial t^2}=A\omega\big[-\cos(kx-\omega t)+\sin(kx-\omega t)\big]$.

The report $\dfrac{\partial^2\psi_3}{\partial t^2}\Big/\dfrac{\partial^2\psi_3}{\partial x^2}=\dfrac{k}{\omega}$ represents a velocity, and then the function Ψ_3

verify the wave equation.

d) $\dfrac{\partial^2\psi_4}{\partial x^2}=ABk\cos(\omega t)\cos(kx)$ and $\dfrac{\partial^2\psi_4}{\partial t^2}=-AB\omega\sin(kx)\sin(\omega t)$.

The report $\dfrac{\partial^2\psi_4}{\partial t^2}\Big/\dfrac{\partial^2\psi_4}{\partial x^2}=\dfrac{-AB\omega\sin(k\,x)\sin(\omega t)}{ABk\cos(\omega t)\cos(k\,x)}$ does not represent a

speed, as consequence the function Ψ_4 does not verify the wave equation.

2) Traveling (progressive) wave, it is a wave propagating in the direction of positive x (with a delay) or negative x (in advance). This is the case for transmitted waves (lagging) and reflected waves (leading).

- The progressive wave propagates in the negative direction is also called a regressive wave.

- Standing wave, these are waves where time plays no role.

Exercise 3.2

$$\dot{x}(t) = -a\omega[\sin(\omega t) + 2\sin(2\omega t)] \Rightarrow \dot{x}_{max} \approx 2.73 a\omega = 136.5 \,(\text{m/s}).$$

Exercise 3.3

We have the function F is non-zero for $\begin{cases} -1 \prec x \prec 1 \text{ at } t = 1\,(s) \\ 2 \prec x \prec 4 \text{ at } t = 2\,(s) \end{cases}$.

Also the function G is non-zero for $\begin{cases} -1 \prec x \prec 1 \text{ at } t = 1\,(s) \\ -4 \prec x \prec -2 \text{ at } t = 2\,(s) \end{cases}$.

The graph of the result $F+G$ is

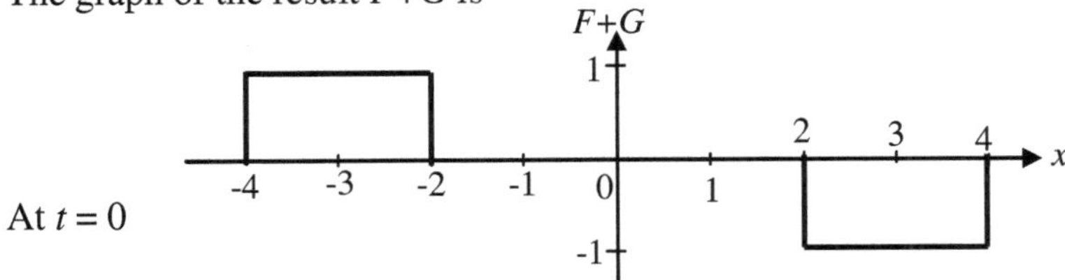

At $t = 0$

At $t = 1$ (s)
(Destructive interference)

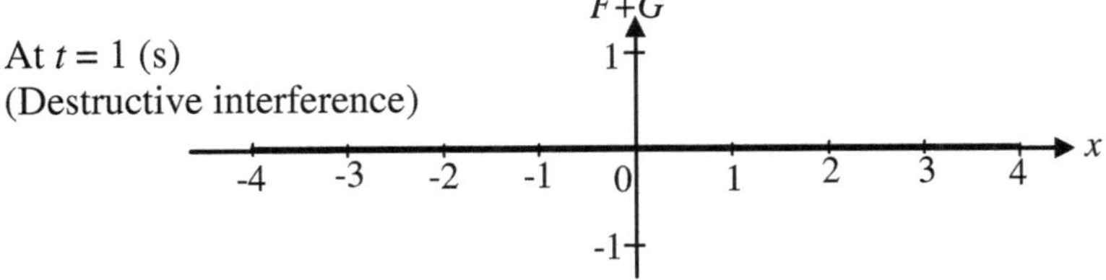

At $t = 2$ (s)

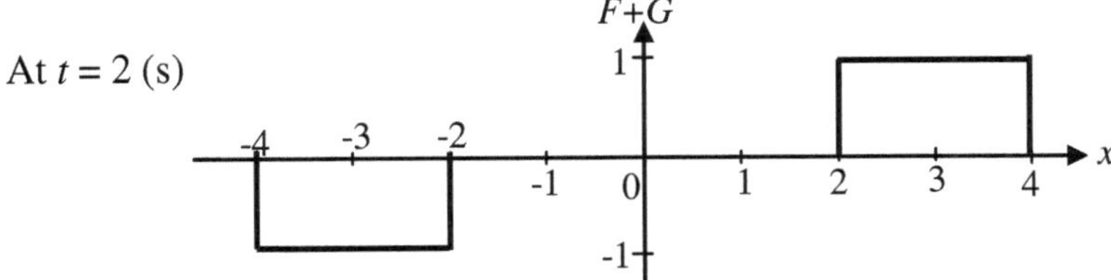

Exercise 3.4

1) The function $\psi(x,t) = \psi_0 \sin[2\pi(50t - 2x)] = \psi_0 \sin(100\pi t - 4\pi x)$.

By comparing it to the general form $\psi = \psi_0 \sin(\omega t - k x)$, we deduce that $k = 4\pi$ and $\omega = 100\pi$. The requested quantities are calculated as follows

The period $T = \dfrac{2\pi}{\omega} = \dfrac{2\pi}{100\pi} = 0.02\,(s).$

The frequency $f = \dfrac{1}{T} = 50\,(\text{Hz})$.

The wavelength $\lambda = \dfrac{2\pi}{k} = \dfrac{2\pi}{4\pi} = 0.5\,(\text{m})$.

The velocity $v = \dfrac{\omega}{k} = \dfrac{2\pi \times 50}{4\pi} = 25\,(\text{m/s})$.

2) The wave described by the function is progressive and the propagation takes place in the direction of increasing x ($x > 0$).

The phase $\varphi = cst$, at fixed $t \Rightarrow \omega t - kx = cst$, so $k\,x = cst$.

$\Rightarrow$ The wave-front is a plane parallel to the (yOz) plane and intersects the x-axis at the point $x = \dfrac{cst}{k}$.

3) At a fixed time t

$$\Delta\varphi = \varphi_1 - \varphi_2 = \dfrac{\pi}{3} \Rightarrow (\omega t - k\,x_1) - (\omega t - k\,x_2) = \dfrac{\pi}{3},$$

$$k(x_2 - x_2) = \dfrac{\pi}{3} \Rightarrow \Delta x = \dfrac{\pi}{3} \times \dfrac{1}{k} = \dfrac{\pi}{3 \times 4\pi} \approx 0.083\,(\text{m}).$$

4) Evaluating of $\Delta\varphi$ for $\Delta t = 1$ (ms)

We take two photos at a single point at two different times, with a time interval $\Delta t = 1$ (ms),

$$\Delta\varphi = (\omega t_1 - k\,x) - (\omega t_2 - k\,x) = \omega \Delta t,$$

Application: $\Delta\varphi = 100\pi \times 10^{-3} = \dfrac{\pi}{10}$ (rd).

Exercise 3.5

The equiphase surfaces are given by $\varphi = cst$, at a fixed time t

i) For ψ_1, $\varphi = \omega t - \vec{k}\vec{r} = \omega t - k_x x - k_y y - k_z z = cst$, with $\vec{k}\begin{pmatrix} k_x \\ k_y \\ k_z \end{pmatrix}$ and $\vec{r}\begin{pmatrix} x \\ y \\ z \end{pmatrix}$.

$\Rightarrow k_x x + k_y y + k_z z = cst$ (it represents the equation of a plane which intersects the three axes at the points $(\dfrac{cst}{k_x}, \dfrac{cst}{k_y}, \dfrac{cst}{k_z})$)).

ii) For ψ_2, $\varphi = \omega t - k r = cst$.

$\Rightarrow k r = cst$ (the geometric locus is a sphere of radius $r = cst$).

Exercise 3.6

We choose a monochromatic wave of the form $\psi = \psi_0 e^{j(kx - \omega t)}$, the dispersion relationships are

i) $\omega = \sqrt{\dfrac{b}{a}}k$, the propagation is non-dispersive, because the phase velocity

$\sqrt{\dfrac{b}{a}}$ does not depend on the pulsation or the wave vector.

ii) $\omega^2 = c^2 k^2 (1 - \alpha k^2)$, the propagation is dispersive.

iii) $\dfrac{\omega^2}{c^2} - k^2 + j\beta\omega = 0$, the propagation is dispersive. Furthermore, the dispersion relationship is complex.

Exercise 3.7
The linear mass of the rope

We have: $\rho = \dfrac{m}{V = S \times l} \Rightarrow \dfrac{m}{l} = \mu = \rho S$, and the section $S = \pi R^2 = \pi(\dfrac{D^2}{4})$.

$\mu = \rho \times \pi \dfrac{D^2}{4} = 0.0558$ (kg/m).

The velocity of the transverse waves is $v = \sqrt{\dfrac{T}{\mu}} = \sqrt{\dfrac{900}{0.0558}} = 127$ (m/s).

Exercise 3.8

1) We note two progressive waves
     - one propagates towards increasing x ($x > 0$),
     - the other propagates towards decreasing x ($x < 0$).
The equation of motion is of the form

$$u(x,t) = A\sin[\omega(t - \frac{x}{v})] + B\sin[\omega(t + \frac{x}{v})].$$

If we take into account the boundary conditions, we obtain

$$u(0,t) = A\sin[\omega t] + B\sin[\omega t] = 0 \Rightarrow A = -B.$$

$$u(L,t) = A\left\{\sin[\omega(t - \frac{L}{v})] - \sin[\omega(t + \frac{L}{v})]\right\} = 2A\sin(\frac{\omega L}{v})\cos(\omega t) = 0,$$

It is void if $\dfrac{\omega L}{v} = n\pi \Rightarrow \omega_n = \dfrac{n\pi v}{L}$ (own pulsations), $n \in \aleph^*$.

The own frequencies $f_n = \dfrac{\omega_n}{2\pi} = \dfrac{n v}{2L} \Rightarrow \lambda_n = \dfrac{v}{f_n}$ (wavelengths).

2) The first three harmonics

i) We have $f_1 = \dfrac{v}{2L}$ and $\lambda_1 = \dfrac{v}{f_1} = 2L \Rightarrow L = \dfrac{\lambda_1}{2}$.

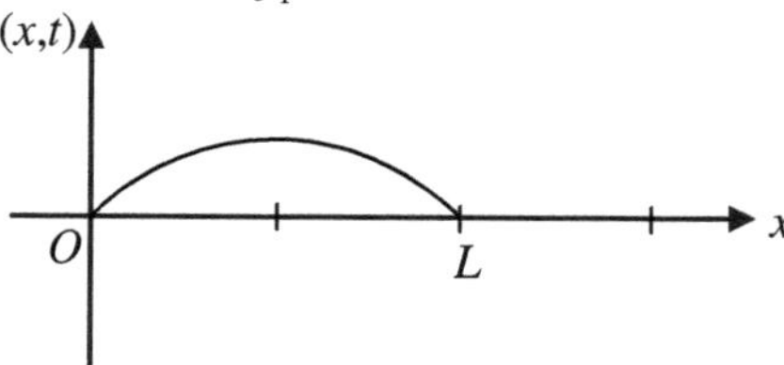

ii) We have $f_2 = \dfrac{2v}{2L}$ and $\lambda_2 = \dfrac{v}{f_2} = L \Rightarrow L = \lambda_2$.

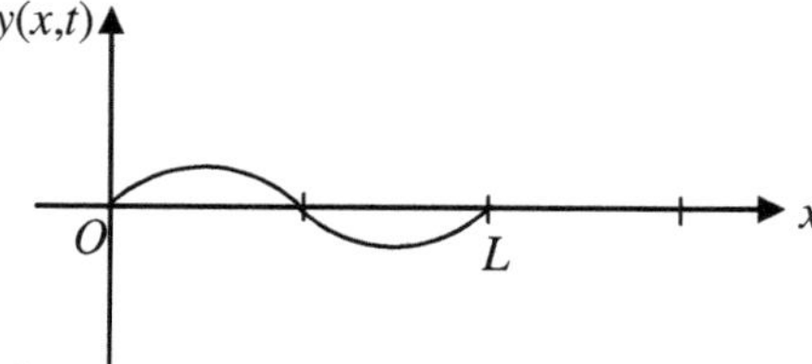

iii) We have $f_3 = \dfrac{3v}{2L}$ and $\lambda_3 = \dfrac{v}{f_3} = \dfrac{2}{3}L \Rightarrow L = \dfrac{3}{2}\lambda_3$.

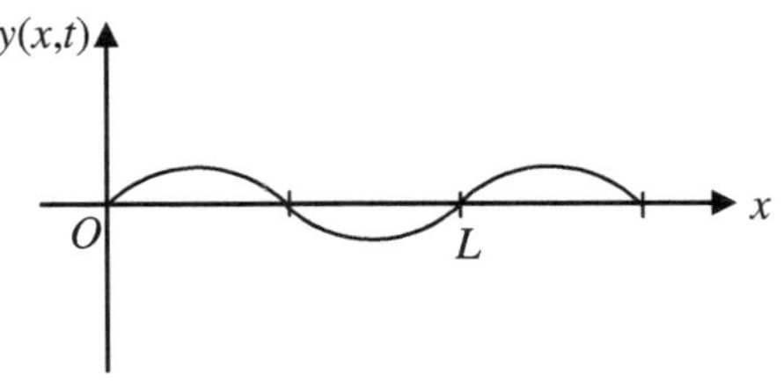

Exercise 3.9

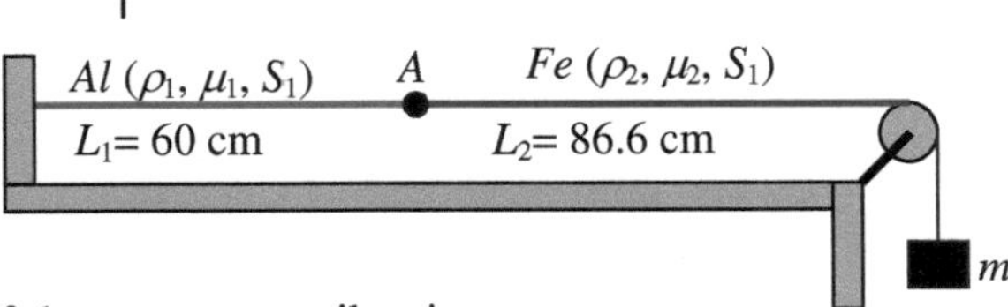

1) The frequency of the transverse vibration

a) - <u>Between the end of *Al* and point *A*</u>

There is an incident wave and a reflected wave (on L_1), its equation is

$$U_1(x,t) = A\sin[\omega(t - \frac{x}{v_1})] + B\sin[\omega(t + \frac{x}{v_1})], \text{ with } v_1 = \sqrt{\frac{T}{\mu_1}}.$$

Point *A* is a node $\Rightarrow U_1(0,t) = U_1(L_1,t) = 0$.

$$U_1(0,t) = A\sin[\omega t] + B\sin[\omega t] = 0 \Rightarrow A = -B.$$

$$U_1(L_1,t) = A\left\{\sin[\omega(t - \frac{L_1}{v_1})] - \sin[\omega(t + \frac{L_1}{v_1})]\right\} = 2A\sin(\frac{\omega L_1}{v_1})\cos(\omega t) = 0,$$

$$\Rightarrow \frac{\omega L_1}{v_1} = n_1\pi \Rightarrow \omega_n = \frac{n_1\pi v_1}{L_1} \text{ (own pulsations)}, n_1 \in \mathbb{N}^*.$$

Own frequencies $f_n = \dfrac{\omega_n}{2\pi} = \dfrac{n_1 v_1}{2L_1} \Rightarrow \lambda_{n1} = \dfrac{v_1}{f_n}$ (wavelengths).

$$\lambda_{n1} = \frac{v_1}{f_n} = 2L_1 \Rightarrow L_1 = n_1\frac{\lambda_{n1}}{2}.$$

b) - <u>Between point *A* and the end of *Fe*</u>

There is a transmitted wave in L_2. Its equation is $U_2(x,t) = C\sin[\omega(t - \frac{x}{v_2})]$,

with $v_2 = \sqrt{\dfrac{T}{\mu_2}}$.

$$U_2(L_2,t) = 0 \Rightarrow \frac{\omega L_2}{v_2} = n_2\pi, \text{ so, } f_n = \frac{\omega_n}{2\pi} = \frac{n_2 v_2}{2L_2}.$$

$$\Rightarrow L_2 = n_2 \frac{\lambda_{n2}}{2}, \text{ with } n_2 \in \aleph^*. \text{ And } \lambda_{n2} = \frac{v_2}{f_n} = 2L_2 \Rightarrow L_2 = n_2 \frac{\lambda_{n2}}{2}.$$

In summary, $\begin{cases} f_n = \dfrac{n_1\, v_1}{2L_1} = \dfrac{n_1}{2L_1}\sqrt{\dfrac{T}{\mu_1}} \\[3mm] f_n = \dfrac{n_2\, v_2}{2L_2} = \dfrac{n_2}{2L_2}\sqrt{\dfrac{T}{\mu_2}} \end{cases}$, the frequency is the same in the wires.

We write $f_n = \dfrac{n_1\, v_1}{2L_1} = \dfrac{n_2\, v_2}{2L_2} \Rightarrow$ the ratio $\dfrac{n_1\, v_1\, L_2}{n_2 v_2 L_1} = 1$.

Which leads to $\dfrac{n_1}{n_2} = \dfrac{v_2 L_1}{v_1 L_2} = \dfrac{L_1}{L_2}\sqrt{\dfrac{\mu_1}{\mu_2}}$.

- Calculation of the linear masses of two wires (the linear mass $\mu = \dfrac{m}{L}$),

The mass is also expressed as a function of the volumic mass by the relation: $m = \rho V = \rho S L$. We use this relationship, we end up with

For the aluminum wire $\mu_1 = \dfrac{m_{Al}}{L_1} = \dfrac{\rho_{Al} S_1 L_1}{L1} = \rho_{Al} S_1$.

The same goes for the iron wire $\mu_2 = \dfrac{m_{Fe}}{L_2} = \dfrac{\rho_{Fe} S_1 L_2}{L_2} = \rho_{Fe} S_1$.

We replace in the previous ratio $\dfrac{n_1}{n_2} = \dfrac{L_1}{L_2}\sqrt{\dfrac{\rho_{Al}}{\rho_{Fe}}} = \dfrac{0.6}{0.86}\sqrt{\dfrac{2.6}{7.8}}$.

Application: $\dfrac{n_1}{n_2} = 0.4 \Rightarrow \begin{cases} n_1 = 2 \\ n_2 = 5 \end{cases}$, and the frequency is $f_n = \dfrac{n_1\, v_1}{2L_1}$.

The velocity v_1 is $v_1 = \sqrt{\dfrac{T}{\rho_{Al}/L_1}} = \sqrt{\dfrac{10\times 9.81}{(2.6\times 10^{-3})/60}} = 1504.6 \ (m/s)$.

The frequency $f_n = \dfrac{2\times 1504.6}{2\times 0.6} = 2507.66 \,(Hz)$.

2) Let's find the number of nodes

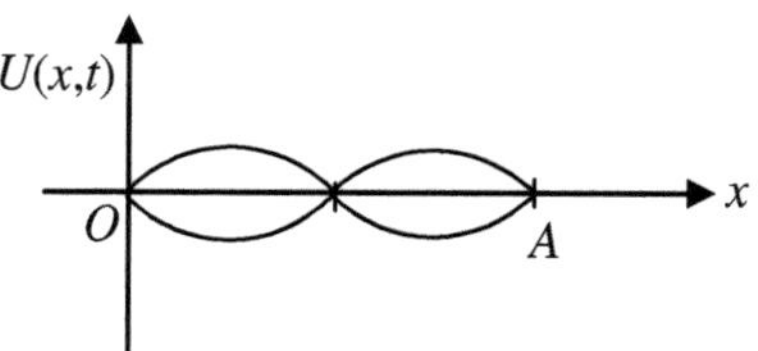

We have $\begin{cases} n_1 = 2 \\ n_2 = 5 \end{cases} \Rightarrow n_1 = 2$ and $\lambda = \dfrac{L_1}{2}$.

- _In the presence of friction_ the number of nodes is $N = 6$ (both ends are not counted).

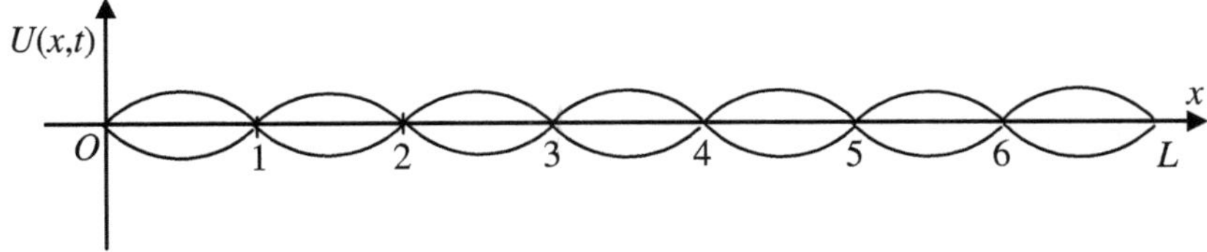

- _No friction_ the number of nodes is $N = 1$ (without the ends).

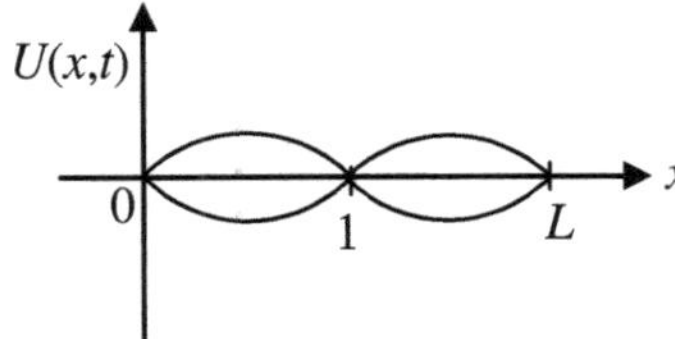

Exercise 3.10

1) We take stock of the forces acting on a strand of rope

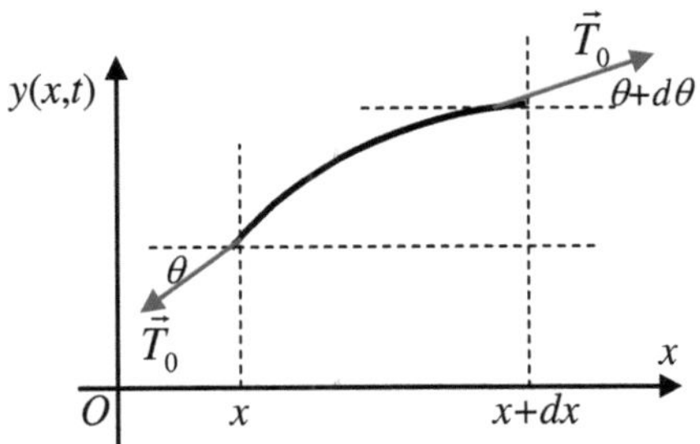

The application of the fundamental principle of dynamics

$$\sum \vec{F} = \vec{T}_x + \vec{T}_{x+dx} = m\vec{a}$$

Projection $\begin{cases} T_0 \cos(\theta + d\theta) - T_0 \cos(\theta) = 0 \\ T_0 \sin(\theta + d\theta) - T_0 \sin(\theta) = ma \end{cases}$

There is movement along the y-axis

$$T_0 \sin(\theta + d\theta) - T_0 \sin(\theta) = m\frac{\partial^2 y(x,t)}{\partial t^2}.$$

Due to low curvature $\Rightarrow \sin(\theta) = \tan(\theta) = \theta$.

Therefore, $T_0 \left[\tan(\theta + d\theta) - \tan(\theta)\right] = m\dfrac{\partial^2 y(x,t)}{\partial t^2}$.

$T_0\, d\left[\tan(\theta)\right] = m\dfrac{\partial^2 y(x,t)}{\partial t^2}$.

In addition, $\tan(\theta) = \dfrac{\partial y}{\partial x} \Rightarrow \dfrac{\partial}{\partial x}\left(\dfrac{\partial y}{\partial x}\right) = \dfrac{\partial^2 y}{\partial x^2} = \dfrac{d}{dx}\left(\tan(\theta)\right)$.

In other hand, $\left(\tan(\theta)\right)' = d\left(\tan(\theta)\right) = \dfrac{\partial^2 y}{\partial x^2}\,dx$ (a total differential),

$T_0\, d\left[\tan(\theta)\right] = T_0\,\dfrac{\partial^2 y(x,t)}{\partial x^2}\,dx = m\dfrac{\partial^2 y(x,t)}{\partial t^2} = \mu\,dx\,\dfrac{\partial^2 y(x,t)}{\partial t^2}$.

The wave propagation equation is $\dfrac{\partial^2 y(x,t)}{\partial t^2} = \dfrac{T_0}{\mu}\dfrac{\partial^2 y(x,t)}{\partial x^2}$, with $\dfrac{T_0}{\mu} = v^2$.

It is in the form $\dfrac{\partial^2 y(x,t)}{\partial t^2} = v^2\dfrac{\partial^2 y(x,t)}{\partial x^2}$.

2) Let us verify that the solution of the equation is $y(x,t) = F(x)f(t)$

The derivative gives $\begin{cases} \dfrac{\partial^2 y(x,t)}{\partial t^2} = F(x)\dfrac{\partial^2 f(t)}{\partial t^2} \equiv F(x)f''(t) \\[2ex] \dfrac{\partial^2 y(x,t)}{\partial x^2} = f(t)\dfrac{\partial^2 F(x)}{\partial x^2} \equiv f(t)F''(x) \end{cases}$.

The replacement in the equation of motion allows us to have

$F(x)f''(t) = v^2 f(t)F''(x) \Rightarrow \dfrac{f''(t)}{f(t)} = v^2\dfrac{F''(x)}{F(x)}$ (differential equation with

two separable variables which only admits a solution if

$\dfrac{f''(t)}{f(t)} = v^2\dfrac{F''(x)}{F(x)} = C \equiv cst$. Which give $\begin{cases} \ddot{F} - \dfrac{C}{v^2}F = 0 \\[2ex] \ddot{f} - C\,f = 0 \end{cases}$.

(Mathematically, the constant C can be positive, negative or zero).
A single value of C corresponds to a physical solution (it is given by the case where $C < 0$).

We pose $C = -\omega^2$. Then, we replace in the system of two differential equations, we arrive at

$$\begin{cases} \ddot{F} + \dfrac{\omega^2}{v^2} F = 0 \\[2mm] \ddot{f} + \omega^2\, f = 0 \end{cases}$$

(classical equations, which we are used to solving).

Their solutions are

$$\begin{cases} F(x) = A_1\, \sin(\omega t) + B_1\, \cos(\omega t) \\[2mm] f(t) = A_2\, \sin(\dfrac{\omega}{v} x) + B_2\, \cos(\dfrac{\omega}{v} x) \end{cases}$$

, with A_1, A_2, B_1 and B_2 are constants to be determined by the initial conditions.

The overall solution of the wave equation is

$$y(x,t) = F(x)f(t) = \left(A_1\, \sin(\omega t) + B_1\, \cos(\omega t)\right) \times \left(A_2\, \sin(\dfrac{\omega}{v} x) + B_2\, \cos(\dfrac{\omega}{v} x) \right)$$

Conditions to the limits $y(0,t) = y(L,t) = 0$,

$y(0,t) = \left(A_1\, \sin(\omega t) + B_1\, \cos(\omega t)\right) B_2 \Rightarrow B_2 = 0$.

Taking into account $B_2 = 0$

$$\Rightarrow y(L,t) = \left(A_1\, \sin(\omega t) + B_1\, \cos(\omega t)\right) \times \left(A_2\, \sin(\dfrac{\omega}{v} L) \right) = 0,$$

Which necessarily implies $\sin(\dfrac{\omega}{v} L)$.

The equation is verified if $\dfrac{\omega}{v} L = n\pi$, $(n \in \aleph^*)$.

$$\Rightarrow \omega_n = \dfrac{n\pi v}{L}, \; n \text{ indicates the number of modes.}$$

The general solution of the wave equation is a superposition of the n vibration modes.

$$y(x,t) = \sum_{n=1}^{\infty}\left(A_1\, \sin(\dfrac{n\pi v}{L} t) + B_1\, \cos(\dfrac{n\pi v}{L} t) \right) \times \left(A_2\, \sin(\dfrac{n\pi v}{L} x) \right).$$

3) In this part, $y(x,0) = 0 \Rightarrow y(x,0) = F(x)$.

$$\begin{cases} y(x,0) = \dfrac{2n}{L} x, & 0 \le x \le \dfrac{L}{2} \\[3mm] y(x,0) = \dfrac{2h}{L}(L-x), & \dfrac{L}{2} \le x \le L. \end{cases}$$

We use Fourier series to write $y(x,t)$.

$$y(x,t) = a_0 + \sum_{n=1}^{\infty} a_n cons(n\omega t) + \sum_{n=1}^{\infty} b_n \sin(n\omega t).$$

Reminder

In general, we calculate the Fourier coefficients of a function $g(x)$, using the

following expressions $\begin{cases} a_0 = \dfrac{1}{T}\int_0^T g(x)dx \\[2mm] a_n = \dfrac{2}{T}\int_0^T g(x)\cos(n\omega\, x)dx \\[2mm] b_n = \dfrac{2}{T}\int_0^T g(x)\sin(n\omega\, x)dx \end{cases}$

We can also use the c_n

$$c_n^2 = a_n^2 + b_n^2 \text{ and } \varphi_n = -artg\left[\frac{a_{bn}}{a_n}\right] \Rightarrow g(x) = a_0 + \sum_{n=1}^{\infty} c_n \cos(n\omega t + \varphi_n).$$

We repeat the exercise; the previous solution $y(x,t)$ is put in the form

$$y(x,t) = \sum_n^{\infty} \left(c_n \sin(\frac{n\pi v}{L}t) + a_n \cos(\frac{n\pi v}{L}t) \right) \sin(\frac{n\pi v}{L} x)$$

$$= \sum_n^{\infty} \sin(\frac{n\pi v}{L} x)\left(c_n \frac{n\pi v}{L}\cos(\frac{n\pi v}{L}t) - \underbrace{a_n \frac{n\pi v}{L}\sin(\frac{n\pi v}{L}t)}_{term=0} \right)$$

According to the data $y(x,0) = 0 \Rightarrow \sum_n^{\infty}\sin(\frac{n\pi v}{L} x)\left(c_n \frac{n\pi v}{L} \right) = 0 \Rightarrow c_n = 0.$

Taking into account $c_n = 0$

$$\Rightarrow y(x,t) = \sum_n^{\infty} a_n \cos(\frac{n\pi v}{L}t) \times \sin(\frac{n\pi}{L} x) + ...$$

The calculus of $a_n = 0$ is done taking into account the initial conditions.

$$y(x,0) = F(x) \Rightarrow \sum_n^{\infty} a_n \sin(\frac{n\pi}{L} x) = F(x).$$

Then, continue the calculation in a manner similar to the exercises in series n° 1.

Exercise 3.11

Non-dispersive medium $\Rightarrow \omega = cst.$

1) We pose $u = t - \dfrac{x}{v}$, then, $\dfrac{\partial u}{\partial t} = 1$ and $\dfrac{\partial u}{\partial x} = -\dfrac{1}{v}$.

In relation to time

The first derivative $\dfrac{\partial \psi}{\partial t} = \dfrac{\partial \psi}{\partial u} \times \underbrace{\dfrac{\partial u}{\partial t}}_{=1} = \dfrac{\partial \psi}{\partial u}$.

The second derivative $\dfrac{\partial^2 \psi}{\partial t^2} = \dfrac{\partial}{\partial t}\left(\dfrac{\partial \psi}{\partial t}\right) = \dfrac{\partial}{\partial t}\left(\dfrac{\partial \psi}{\partial u}\right) = \dfrac{\partial^2 \psi}{\partial u^2} \times \underbrace{\dfrac{\partial u}{\partial t}}_{=1} = \dfrac{\partial^2 \psi}{\partial u^2}$ (1).

In relation to space

The first derivative $\dfrac{\partial \psi}{\partial x} = \dfrac{\partial \psi}{\partial u} \times \underbrace{\dfrac{\partial u}{\partial x}}_{=-1/v} = \dfrac{\partial \psi}{\partial u} \times \left(\dfrac{-1}{v}\right) = -\dfrac{\partial \psi}{v\,\partial u}$.

The second derivative

$$\dfrac{\partial^2 \psi}{\partial x^2} = \dfrac{\partial}{\partial x}\left(\dfrac{\partial \psi}{\partial x}\right) = \dfrac{\partial}{\partial x}\left(-\dfrac{\partial \psi}{v\,\partial u}\right) = \dfrac{\partial}{\partial u}\left(\dfrac{-\partial \psi}{v\,\partial u}\right) \times \underbrace{\dfrac{\partial u}{\partial x}}_{=-1/v} = \dfrac{\partial^2 \psi}{v^2 \partial u^2}$$ (2).

From the equation (1) $\Rightarrow \dfrac{\partial^2 \psi}{\partial u^2} = \dfrac{\partial^2 \psi}{\partial t^2}$ (3)

From the equation (2) $\Rightarrow \dfrac{\partial^2 \psi}{\partial u^2} = v^2 \dfrac{\partial^2 \psi}{\partial x^2}$ (4)

We have (3) = (4) $\Rightarrow \dfrac{\partial^2 \psi}{\partial t^2} - v^2 \dfrac{\partial^2 \psi}{\partial x^2} = 0$.

2) A solution in the form: $\psi = F(x).f(t)$

See the answer to question 2 of the previous exercise.

Exercise 3.12

1) The expression $\psi(x,t)$ of a given point of the rope

Generally speaking, the propagation equation is given in the form

$\psi(x,t) = A\cos(\omega t - \vec{k}.\vec{r})$, or $\psi(x,t) = A\sin(\omega t - \vec{k}.\vec{r})$.

In this exercise, $\omega = 2\pi f = 12\pi$ (rd/s); $A = a = 3$ (mm) $= 3\ 10^{-3}$ (m), the

velocity $c = 24$ (m/s) $\Rightarrow$ The wave vector: $k = \dfrac{\omega}{c} = \dfrac{12\pi}{24} = \dfrac{\pi}{2}$ (m^{-1}).

$\psi(0,0) = 0 \Rightarrow$ This means that the elongation equation is expressed in sine

$\psi(x,t) = A\sin(\omega t - \vec{k}.\vec{r})$. Taking into account the data

$\psi(x,t) = 3 \ 10^{-3} \sin[12\pi t - (-\dfrac{\pi}{2}x)]$, (the sign $-$ means negative direction).

- Let us show that the wave admits two solutions

We notice that the function of the elongation verifies the wave equation

$\dfrac{\partial^2 \psi}{\partial x^2} - \dfrac{1}{v^2}\dfrac{\partial^2 \psi}{\partial t^2} = 0$, the latter admits two solutions, which are

$$\psi(x,t) = \underbrace{f(t+\dfrac{x}{v})}_{propagatia \ x \prec 0} + \underbrace{g(t-\dfrac{x}{v})}_{propagatia \ x \succ 0} \ .$$

2) We choose solutions such as $x = 0.5$ (m) $\Rightarrow \psi(0.5,0) \succ 0$,

The displacement of the point is given by

$\psi(0.5,t) = 3 \ 10^{-3} \sin[0 + \dfrac{\pi}{2}\times 0.5)] = 2.1 \ 10^{-3} \succ 0$ (positive).

3) The velocity expression

$v(x,t) = \dot{\psi}(x,t) = \dfrac{\partial \psi(x,t)}{\partial t} = 36\pi \ 10^{-3} \cos(12\pi t + \dfrac{\pi}{2}x)$.

Exercise 3.13

1) The form of elongation

$S_1(x,t)$ is of the form $f(t - \dfrac{x}{v}) \Rightarrow$ propagation towards increasing x.

$S_2(x,t)$ is of the form $f(t + \dfrac{x}{v}) \Rightarrow$ propagation towards decreasing x.

2) The result wave

$$S(x,t) = S_1 + S_2 = 2S_0 \cos[k\,x + (\dfrac{\varphi_1 - \varphi_2}{2})].\cos[\omega t - (\dfrac{\varphi_1 - \varphi_2}{2})],$$

We pose $F(x) = 2S_0 \cos[k\,x + (\dfrac{\varphi_1 - \varphi_2}{2})]$.

a) The points where the function $F(x)$ is minimal are obtained when

$S(x,t) = 0$, $\forall\ t$, which correspond to: $\cos[k\,x + (\dfrac{\varphi_1 - \varphi_2}{2})] = 0$.

Either $kx = \dfrac{\pi}{2} + n\pi - (\dfrac{\varphi_1 - \varphi_2}{2})$, $n \in Z$ (relative integer).

Since $k = \dfrac{2\pi}{\lambda} \Rightarrow x = \dfrac{\lambda}{4} + n\dfrac{\lambda}{2} - \dfrac{\lambda}{2\pi}(\dfrac{\varphi_1 - \varphi_2}{2})$.

The points of zero amplitude, of abscissa x are such that

$$x_N = \frac{\lambda}{4} + n\frac{\lambda}{2} - \frac{\lambda}{2\pi}(\frac{\varphi_1 - \varphi_2}{2}).$$

- The distance between two consecutive points is

$$x_N(n+1) - x_N(n) = \frac{\lambda}{2},$$ these points represent the points of minimum

vibration, they are called vibration nodes.

b) Let's determine the points where $|S(x,t)|$ is maximum

$$S(x,t) \text{ is max, if } \cos[k\,x + (\frac{\varphi_1 - \varphi_2}{2})] = \pm 1.$$

That's to say $kx + (\frac{\varphi_1 - \varphi_2}{2}) = n\pi$, $n \in Z$ (relative integer).

Then, $x = n\frac{\lambda}{2} - \frac{\lambda}{2\pi}(\frac{\varphi_1 - \varphi_2}{2}).$

The points of maximum amplitude, of abscissa x are such that

$$x_V = n\frac{\lambda}{2} - \frac{\lambda}{2\pi}(\frac{\varphi_1 - \varphi_2}{2}).$$

The distance between two consecutive points is

$$x_V(n+1) - x_V(n) = \frac{\lambda}{2} \text{ (They are called vibration bellies).}$$

Exercise 3.14

1) If $u(x,t)$ expresses the transverse displacement of a very small segment of the rope, the application of the fundamental principle of dynamics to the latter leads to the differential equation of motion

$$\frac{\partial^2 u(x,t)}{\partial x^2} - \frac{1}{v^2}\frac{\partial^2 u(x,t)}{\partial t^2} = 0 \text{ (named wave equation) and the velocity } v = \sqrt{\frac{T_0}{\mu}}$$

2) By definition, during stationary vibrations of a rope, a point of it will carry out transverse oscillations whose maximum amplitude is independent of that of other neighboring points and of time.

    - It only depends on its position x along the string and the initial conditions of the movement: we denote this amplitude by $f(x)$.

    - If we assume these oscillations are perfectly harmonic in time, we can write for a point and, for a given mode or pulsation ω

$$f(x) = A\sin(k\,x) + B\cos(k\,x) \text{ (standing waves).}$$

By injecting the expression of $u(x,t)$ into the previous equation, we find that
$$\frac{\partial^2 f(x)}{\partial x^2} + \frac{\omega^2}{v^2} f(x) = 0.$$

Whose solution will be of the form $f(x) = A\sin(k\,x) + B\cos(k\,x)$, by posing $\frac{\omega}{v} = k = \frac{2\pi}{\lambda}$, which means that for a homogeneous string the dispersion relation is linear $\omega = k\,v$.

3) For $u(0,t) = u(l,t) = 0$ and for a given mode ω

$\Rightarrow B = 0$ and $\omega_n = k_n\, v = \frac{n\pi v}{l}$, where n is an integer.

Only oscillations of determined pulsations can arise in a rope fixed at its ends. Consequently, $u_n(x,t) = A_n \sin(k_n x)\cos(\omega_n t + \varphi_n)$.

4) For a homogeneous rope, a more general movement will be written
$$u(x,t) = \sum_n A_n \sin(k_n x)\cos(\omega_n t + \varphi_n).$$

5) If at $t = 0$, $\dot{u}(x,t) = 0 \; \forall \, x \; \Rightarrow \; u(x,t) = \sum_n A_n \sin(k_n x)\cos(\omega_n t)$.

6) If more, at $t = 0$, $f(x) = a\sin(\frac{3\pi}{l} x)$, then $A_n = 0$ and this $\forall \, n \neq 3$ and

$A_3 = a$. Finally, $u(x,t) = a\sin(\frac{3\pi}{l} x)\cos(\frac{3\pi v}{l} t)$.

Exercise 3.15

Application: $\omega_1 \approx 47.9\,(\text{rd/s})$; $\omega_2 \approx 52.1\,(\text{rd/s})$.

The period $T_B = \dfrac{2\pi}{\omega_2 - \omega_1} = 1.5\,(\text{s})$.

Exercise 3.16

1) The phase velocity is $v_\varphi = \dfrac{\omega}{k}$ and the group velocity is $v_g = \dfrac{d\omega}{dk}$.

So, $\omega = v_\varphi \times k$ (we derive this relation with respect to k)

$$\Rightarrow \frac{d\omega}{dk} = v_g = \frac{d}{dk}\left(v_\varphi \times k\right) = v_\varphi + k \times \frac{dv_\varphi}{dk} \quad (1)$$

As $k = \dfrac{2\pi}{\lambda}$, we take *the Napierian logarithm* of both sides

$\Rightarrow \ln(k) = \ln(2\pi) - \ln(\lambda)$, then we drift $\dfrac{dk}{k} = -\dfrac{d\lambda}{\lambda}$.

So, $dk = -(\dfrac{d\lambda}{\lambda})k$, we replace it in equation (1), we obtain

$$v_g = v_\varphi - \lambda \times \frac{dv_\varphi}{d\lambda} \quad \text{(Rayleigh criterion)}.$$

- Particular case: case where $v_\varphi = \dfrac{c}{n} \Rightarrow \dfrac{dv_\varphi}{d\lambda} = \dfrac{dv_\varphi}{dn} \times \dfrac{dn}{d\lambda} = -\dfrac{c}{n^2}\dfrac{dn}{d\lambda}$.

Which give: $v_g = v_\varphi + \dfrac{\lambda c}{n^2} \times \dfrac{dn}{d\lambda}$. Finally, $v_g = v_\varphi\left[1 + \dfrac{\lambda}{n} \times \dfrac{dn}{d\lambda}\right]$.

2) The report $\dfrac{v_g}{v_\varphi} = 1 + \dfrac{\lambda}{n} \times \dfrac{dn}{d\lambda} = 1 + \dfrac{0{,}55 \quad 10^{-6}}{1.64}(-2 \quad 10^{-5}) = 0.94$.

3) we have $f^2 = c^2\left(\dfrac{k^2}{4\pi^2} + \dfrac{b^2}{4a^2}\right)$, and we know that $\omega = 2\pi f$, then,

$$v_\varphi = \frac{\omega}{k} = \frac{2\pi f}{k} = \frac{2\pi c}{k}\sqrt{\frac{k^2}{4\pi^2} + \frac{b^2}{4a^2}}.$$

i) Let's look for v_g based on the given parameters

The group velocity $v_g = \dfrac{d\omega}{dk}$ and $f^2 = c^2\left(\dfrac{k^2}{4\pi^2} + \dfrac{b^2}{4a^2}\right)$.

ii) Let's calculate the product between the two speeds (group and phase)

The derivative $d\left(\dfrac{\omega^2}{4\pi^2}\right) = d\left\{c^2\left[\dfrac{k^2}{4\pi^2} + \dfrac{b^2}{4a^2}\right]\right\} \Leftrightarrow \dfrac{2\omega\, d\omega}{4\pi^2} = c^2\dfrac{2k\,dk}{4\pi^2}$.

So, $\dfrac{d\omega}{dk} = c^2\dfrac{k}{\omega} = c^2\dfrac{1}{v_\varphi} = v_g \Rightarrow v_\varphi \times v_g = c^2$.

- Discussion: on the requirements of relativistic mechanics

$$v_\varphi = \frac{2\pi c}{k}\sqrt{\frac{k^2}{4\pi^2} + \frac{b^2}{4a^2}} = c\sqrt{1 + \frac{b^2\pi^2}{a^2k^2}} \Rightarrow v_\varphi \succ c.$$

$$v_\varphi \times v_g = c^2 \Rightarrow v_\varphi = \frac{c}{\sqrt{1 + \dfrac{b^2\pi^2}{a^2k^2}}} \prec c \Rightarrow v_\varphi \prec c.$$

iii) Case where the medium is non-dispersive

- In the case of a dispersive medium we see that $v_\varphi = \dfrac{\omega}{k}$, with $\omega = cst$ (the wave propagates with a single wavelength) $\Rightarrow v = cst$.

- If the medium is non-dispersive ($n = cst \ \forall \ t$ et $\forall \ x$)

$$v_g = v_\varphi \left[1 + \frac{\lambda}{n} \times \frac{dn}{d\lambda} \right] = v_\varphi \text{ (since } \frac{dn}{d\lambda} = 0 \text{, from the fact that } n = cst).$$

In a non-dispersive medium $v_g = v_\varphi$.

Exercise 3.17

As the electric field of the plane wave is oriented along the x-axis, the condition $\vec{k}.\vec{E} = 0$ imposes that $k_x = 0$ and the vector $\vec{k}$ is in plan (yOz) $\vec{k} = k_y \vec{e}_y + k_z \vec{e}_z$.

To determine the magnetic field, we use the relation $\vec{E} = -\dfrac{\partial \vec{B}}{\partial t}$, which is

written $j(\vec{k} \wedge \vec{E}) = -\dfrac{\partial \vec{B}}{\partial t}$. So, $j\left(\begin{pmatrix} 0 \\ k_y \\ k_z \end{pmatrix} \wedge \begin{pmatrix} E_x \\ 0 \\ 0 \end{pmatrix} \right) = -\dfrac{\partial}{\partial t} \begin{pmatrix} B_x \\ B_y \\ B_z \end{pmatrix} .e^{j(\vec{k}.\vec{r} - \omega t)}$.

The only following y-component is non-zero, we obtain a magnetic field of

the form: $\vec{B} = B_0 e^{j(\vec{k}.\vec{r} - \omega t)} \vec{e}_y$, with $B_0 = -\dfrac{k_x}{E_0} \omega$.

Exercise 3.18

We determine the magnetic field $\vec{B}$ by the relation $\vec{E} = -\dfrac{\partial \vec{B}}{\partial t}$.

The rotational of the electric field is

$$\overrightarrow{rot}(\vec{E}) = E_0 \left[\left(\frac{n\pi}{L_y} \right) \cos\left(\frac{n\pi y}{L_y} \right) \vec{e}_x - jk_g \sin\left(\frac{n\pi y}{L_y} \right) \vec{e}_y \right] e^{j(k_g x - \omega t)}.$$

And after integration with respect to time, we find

$$\vec{B} = -\frac{E_0}{\omega} \left[j(\frac{n\pi}{L_y}) \cos\left(\frac{n\pi y}{L_y} \right) \vec{e}_x + k_g \sin\left(\frac{n\pi y}{L_y} \right) \vec{e}_y \right] e^{j(k_g x - \omega t)}.$$

The magnetic field in the waveguide is not perpendicular to the wave vector if the electric field is transverse.

Exercise 3.19

1) Using the data, we find that $\omega_\tau = 3 \times 10^{-6}$ (rd/s).

2) The dispersion relation is $k = \dfrac{\omega}{v}(1+ j\omega_\tau)^{-0.5}$.

On the other hand, we see that the quantity $\omega_\tau << 1$ (negligible in front 1), we can do limited development to write: $k = \dfrac{\omega}{v} - \dfrac{1}{2} j \dfrac{\omega_\tau^2}{v}$.

3) By injecting this complex wave number into the writing of the pressure wave, we obtain $P(x,t) = P_0\, e^{\omega t} e^{-jk_{re}x} e^{k_{im}x} = P_0\, e^{-x/L} e^{j(\omega t - k_{re}x)}$.

The attenuation length L is by identification $L = \dfrac{2v}{\omega_\tau^2} = \dfrac{2\rho_0 v^3}{\omega^2 \eta}$.

Exercise 3.20

1) We obtain the magnetic field by the relation

$\dfrac{\partial \vec{B}_1}{\partial t} = -\overrightarrow{rot}(\vec{E}) = -j(\vec{k}_1 \wedge \vec{E}_1)$. Which give

$$\dfrac{\partial \vec{B}_1}{\partial t} = -jk \begin{pmatrix} 0 \\ \cos(\theta) \\ \sin(\theta) \end{pmatrix} \wedge \begin{pmatrix} E_0 \\ 0 \\ 0 \end{pmatrix}.e^{j(\vec{k}_1.\vec{r}-\omega t)} = -jk\,\vec{E}_0 \begin{pmatrix} 0 \\ \sin(\theta) \\ -\cos(\theta) \end{pmatrix}.e^{j(\vec{k}_1.\vec{r}-\omega t)},$$

After integration with respect to time (by choosing a zero integration coefficient, because there is no static magnetic field),

We find $\vec{B}_1 = \dfrac{E_0}{v} \begin{pmatrix} 0 \\ \sin(\theta) \\ -\cos(\theta) \end{pmatrix}.e^{j(\vec{k}_1.\vec{r}-\omega t)}$.

A similar calculation allows us to have the vector $\vec{B}_2$

$$\vec{B}_2 = \dfrac{-E_0}{v} \begin{pmatrix} 0 \\ \sin(\theta) \\ \cos(\theta) \end{pmatrix}.e^{j(\vec{k}_2.\vec{r}-\omega t)}.$$

2) The resulting electric field is $\vec{E} = \vec{E}_1 + \vec{E}_2 = E_0\left[e^{j(\vec{k}_1 . \vec{r} - \omega t)} + e^{j(\vec{k}_2 . \vec{r} - \omega t)}\right]\vec{e}_x$,

and its real part is

$$\mathrm{Re}\{\vec{E}\} = E_0\left[\cos[k_y \cos(\theta) + k_z \sin(\theta) - \omega t] + \cos[k_y \cos(\theta) - k_z \sin(\theta) - \omega t]\right]\vec{e}_x$$
$$= E_0\left(\cos[k_y \cos(\theta) - \omega t]\cos[k_z \sin(\theta)]\right).$$

The same method gives for the total magnetic field $\vec{B} = \vec{B}_1 + \vec{B}_2$;

$$\overrightarrow{rot}(\vec{B}) = -\frac{2E_0}{v}\begin{pmatrix} 0 \\ \sin[k_y \cos(\theta) - \omega t] \times \sin[k_z \sin(\theta)] \times \sin(\theta) \\ \cos[k_y \cos(\theta) - \omega t] \times \cos[k_z \sin(\theta)] \times \cos(\theta) \end{pmatrix}.$$

3) The Poynting vector, calculated from the real parts of the electric and magnetic fields, is $\vec{\Pi} = \varepsilon_0 v E_0^2 \begin{pmatrix} 0 \\ 4\cos^2[k_y \cos(\theta) - \omega t] \times \cos^2[k_z \sin(\theta)] \times \cos(\theta) \\ -\sin[2k_y \cos(\theta) - 2\omega t] \times \sin[2k_z \sin(\theta)] \times \sin(\theta) \end{pmatrix}$

4) In time average, the component $\vec{\Pi}_z$ of the Poynting vector is null, because the function $\sin[2k_y \cos(\theta) - 2\omega t]$ has zero average. On the other hand, the component $\vec{\Pi}_y$ is not zero, and we find

$$\vec{\Pi}_y = 2\varepsilon_0 c E_0^2 \cos^2[k_z \sin(\theta)]\cos(\theta)\vec{e}_y.$$

The resulting wave presents an interference structure with interference fringes separated by a distance $\Delta z = \dfrac{\pi}{k\sin(\theta)} = \dfrac{\lambda}{2\sin(\theta)}$.

References

- M. Bendaoud, Vibrations and Waves, courses and exercises: module TP010, OPU (1994).
- Alonso, Finn, General Physics, Volume 2: Fields and waves, 2nd ed., Dunod, Paris, 2005.
- J-P. Mathieu, vibrations and propagation phenomena, Volume I, oscillators, Masson et Cie, Publishers, 1974.
- A. Fouillé, P. Déréthé, Course in physics of vibrations, Dunod, Paris, 1976.
- J. Bruneaux, J. Matricon Vibrations waves, ellipses, 2008.
- T. Bécherrawy, Vibrations and waves, La voiser, 2010.
- H. J. Pain, The Physics of Vibrations and Waves, 6th Edition, Formerly of Department of Physics, Imperial College of Science and Technology, London, UK, 2005.
- F. S. Crawford, Berkeley Physics Course, Volume 3: Waves, Dunod, 1999.
- H. J. Pain, The Physics of Vibrations and Waves, 6th Edition, John Wiley & Sons, Ltd. NY, 2005.